ENGAGING THE WORLD:
THE USE OF EMPIRICAL RESEARCH IN BIOETHICS AND
THE REGULATION OF BIOTECHNOLOGY

Biomedical and Health Research

Volume 58

Recently published in this series:

Vol. 57. A. Nosikov and C. Gudex (Eds.), EUROHIS: Developing Common Instruments for Health Surveys
Vol. 56. P. Chauvin and the Europromed Working Group (Eds.), Prevention and Health Promotion for the Excluded and the Destitute in Europe
Vol. 55. J. Matsoukas and T. Mavromoustakos (Eds.), Drug Discovery and Design: Medical Aspects
Vol. 54. I.M. Shapiro, B.D. Boyan and H.C. Anderson (Eds.), The Growth Plate
Vol. 53. C. Huttin (Ed.), Patient Charges and Decision Making Behaviours of Consumers and Physicians
Vol. 52. J.-F. Stoltz (Ed.), Mechanobiology: Cartilage and Chondrocyte, Vol. 2
Vol. 51. G. Lebeer (Ed.), Ethical Function in Hospital Ethics Committees
Vol. 50. R. Busse, M. Wismar and P.C. Berman (Eds.), The European Union and Health Services
Vol. 49. T. Reilly (Ed.), Musculoskeletal Disorders in Health-Related Occupations
Vol. 48. H. ten Have and R. Janssens (Eds.), Palliative Care in Europe – Concepts and Policies
Vol. 47. H. Aldskogius and J. Fraher (Eds.), Glial Interfaces in the Nervous System – Role in Repair and Plasticity
Vol. 46. I. Philp (Ed.), Family Care of Older People in Europe
Vol. 45. L. Turski, D.D. Schoepp and E.A. Cavalheiro (Eds.), Excitatory Amino Acids: Ten Years Later
Vol. 44. R. Coppo and Dr. L. Peruzzi (Eds.), Moderately Proteinuric IgA Nephropathy in the Young
Vol. 43. B. Shaw, G. Semb, P. Nelson, V. Brattström, K. Mølsted and B. Prahl-Andersen, The Eurocleft Project 1996-2000
Vol. 42. J.-F. Stoltz (Ed.), Mechanobiology: Cartilage and Chondrocyte
Vol. 41. T.M. Gress (Ed.), Molecular Pathogenesis of Pancreatic Cancer
Vol. 40. H. Leino-Kilpi, M. Välimäki, M. Arndt, T. Dassen, M. Gasull, C. Lemonidou, P.A. Scott, G. Bansemir, E. Cabrera, H. Papaevangelou and J. Mc Parland, Patient's Autonomy, Privacy and Informed Consent
Vol. 39. J.-M. Graf von der Schulenburg (Ed.), The Influence of Economic Evaluation Studies on Health Care Decision-Making
Vol. 38. N. Yoganandan and F.A. Pintar (Eds.), Frontiers in Whiplash Trauma
Vol. 37. J.M. Ntambi (Ed.), Adipocyte Biology and Hormone Signaling
Vol. 36. F.F. Parl, Estrogens, Estrogen Receptor and Breast Cancer
Vol. 35. M. Schlaud (Ed.), Comparison and Harmonisation of Denominator Data for Primary Health Care Research in Countries of the European Community
Vol. 34. G.J. Bellingan and G.J. Laurent (Eds.), Acute Lung Injury: From Inflammation to Repair
Vol. 33. H.H. Goebel, S.E. Mole and B.D. Lake (Eds.), The Neuronal Ceroid Lipofuscinoses (Batten Disease)
Vol. 32. B.J. Njio, B. Prahl-Andersen, G. ter Heege, A. Stenvik and R.S. Ireland (Eds.), Quality of Orthodontic Care – A Concept for Collaboration and Responsibilities
Vol. 31. B.J. Njio, A. Stenvik, R.S. Ireland and B. Prahl-Andersen (Eds.), EURO-QUAL
Vol. 30. J.-F. Stoltz, M. Singh and P. Riha, Hemorheology in Practice
Vol. 29. G. Pawelec (Ed.), EUCAMBIS: Immunology and Ageing in Europe
Vol. 28. A.M.N. Gardner and R.H. Fox, The Venous System in Health and Disease
Vol. 27. P.A. Frey and D.B. Northrop (Eds.), Enzymatic Mechanisms
Vol. 26. P. Jenner (Ed.), A Molecular Biology Approach to Parkinson's Disease
Vol. 25. N.J. Gooderham (Ed.), Drug Metabolism: Towards the Next Millennium
Vol. 24. S.S. Baig (Ed.), Cancer Research Supported under BIOMED 1

ISSN 0929-6743

Engaging the World:
The Use of Empirical Research in Bioethics and the Regulation of Biotechnology

Edited by

Søren Holm

*Institute of Medicine, Law and Bioethics, University of Manchester,
United Kingdom*

and

Monique F. Jonas

*Centre for Social Ethics and Policy, University of Manchester,
United Kingdom*

IOS Press

Amsterdam • Berlin • Oxford • Tokyo • Washington, DC

© 2004, The authors mentioned in the table of contents

All rights reserved. No part of this book may be reproduced, stored in a retrieval system, or transmitted, in any form or by any means, without prior written permission from the publisher.

ISBN 1 58603 400 6
Library of Congress Control Number: 2003116097

Publisher
IOS Press
Nieuwe Hemweg 6B
1013 BG Amsterdam
The Netherlands
fax: +31 20 620 3419
e-mail: order@iospress.nl

Distributor in the UK and Ireland
IOS Press/Lavis Marketing
73 Lime Walk
Headington
Oxford OX3 7AD
England
fax: +44 1865 75 0079

Distributor in the USA and Canada
IOS Press, Inc.
5795-G Burke Centre Parkway
Burke, VA 22015
USA
fax: +1 703 323 3668
e-mail: iosbooks@iospress.com

LEGAL NOTICE
The publisher is not responsible for the use which might be made of the following information.

PRINTED IN THE NETHERLANDS

Preface

The papers in this book are all the result of work performed as part of the Empirical Methods in Bioethics (EMPIRE) project sponsored by the European Commission, DG-Research as part of the Quality of Life and Management of Living Resources research program. We thank the Commission for its generous support.

The papers collected here are mainly concerned with one of the central objectives of the EMPIRE project: to investigate the ways in which data generated by empirical research can be relevant to bioethical thinking and regulation concerning health care and human use of biotechnology, but also touch on the issue of what kind of empirical research that is most relevant in this context.

They approach these questions in a number of ways, from the very theoretical to the more practical. Some are skeptical to the use of empirical research and data, whereas others are much more positive. They thereby reflect the full breadth of the analysis carried out as part of the project.

Other papers from the project have been published in thematic issues of the journals Health Care Analysis (2003, Volume 11, Issue 1) and Politeia (2002, Volume 18, Issue 67), and in Häyry, M, Takala, T (Eds.), *Scratching the Surface of Bioethics*, Amsterdam and New York: Rodopi, 2003.

Søren Holm
Centre for Social Ethics and Policy
University of Manchester

Contents

Preface   v

Section 1. What Role Can Empirical Research Play in Bioethics?

A Defence of Empirical Bioethics   3
Søren Holm

Healthcare and the Habermasian Public Sphere   8
Andrew Edgar

Putting Empirical Studies in their Place   18
John Harris

Can There Be Moral Experts?   28
Louise Irving and Nina Hallowell

Section 2. Empirical Bioethics Considered: Critiquing the Use of Empirical Research

What should we do about it? Implications of the Empirical Evidence in Relation to Comprehension and Acceptability of Randomisation   41
Angus Dawson

Use and Abuse of Empirical Knowledge in Contemporary Bioethics: A Critical Analysis of Empirical Arguments Employed in the Controversy Surrounding Stem Cell Research   53
Jan Helge Solbakk

Carol Gilligan and her Different Voice   61
Heta Aleksandra Gylling

Who Should Decide and Why? The Futility of Philosophy, Sociology and Law in Institutionalised Bioethics and the Unwarrantability of Ethics Committees   69
Tuija Takala

Empirical Investigations of Moral Development   76
Michael Norup

Theory and Methodology of Empirical-Ethical Research   89
Lieke Van der Scheer, Ghislaine Van Thiel, Johannes Van Delden and Guy Widdershoven

Section 3. Empirical Bioethics in Practice

An Empirical Study of the Informed Consent Process of a Clinical Trial 99
Anne Gammelgaard

Stimulating Public Debate on the Ethical and Social Issues Raised by the New Genetics 109
Mairi Levitt, Kate Weiner and John Goodacre

The Use of Computer Simulation and Artificial Intelligence in the Study of Ethical Components of Medical Decision-Making 119
Peter Øhrstrøm, Jørgen Albretsen and Søren Holm

Conclusion

Empirical Research in Bioethics: Report for the European Commission 131
Søren Holm and Louise Irving

Author Index 157

Section 1

What Role Can Empirical Research Play in Bioethics?

A Defence of Empirical Bioethics

Søren HOLM
University of Manchester; University of Oslo

The purpose of this project is to look at the use of empirical data in bioethics and in the regulation of "bioethical issues", and in this paper I will try to delineate a number of different kinds of empirical data, to describe a number of different uses of empirical data, and finally to present some arguments for why empirical data are important for a broad range of arguments within bioethics.

Even the briefest reading of the bioethical literature reveals that it is full of references to what people want, what they think, and what effect various public policies will have, and these seem *a priori* all to be empirical matters. So, how could there be a need to defend the use of empirical data, when everyone in the field is implicated in their use?

Well, strange as it may seem, some of the people who use empirical data, in other writings claim that some use is illegitimate. Before we can assess this claim we will, however, have to make a few initial distinctions.

First we need to be clear about what kind of empirical data we are talking about, because different kinds of empirical data raise different kinds of problems. The first kind of data which is used regularly, and I think in most cases quite unproblematically, are biological data about human beings and other organisms. Many bioethical arguments do, for instance, implicitly rely on the premise that human beings are mortal. This is clearly not a conceptual truth, there is no inherent contradiction in the concept of an immortal human being, but nevertheless people usually do not protest about the use of empirical data of this kind.

The second kind of data we need to look at are what could broadly be called psychological data. Data about human thought processes, wishes, desires, beliefs, attitudes, etc. It is often claimed that such data are irrelevant to ethics, especially beliefs and attitudes, because the fact that most or even all people believe something to be wrong does not make it wrong. I believe this to be obviously true, but as I will try to show later that this does not close the book on the relevance of psychological data.

The third kind of data are data of a broadly sociological nature. Data about social interactions, health care systems, kinship systems etc. These have also claimed to be of no relevance to bioethics, usually because they are claimed to be easily malleable.

In the following I will mainly concentrate on empirical data of a psychological and sociological nature, since this is where the controversy is most distinct.

I will, however, need to make one more initial clarification before we can move on. The philosophical user of empirical data might claim something like the following: "You have completely misunderstood me. When I talk about what people think or desire I am not drawing on any kind of empirical psychology, I am drawing on philosophical psychology, which is something completely different. The "philosophical psychologist" does not need any empirical data to work on. It is all conceptual."

It is undoubtedly true that empirical psychology and philosophical psychology are two very different enterprises, but also undoubtedly false that philosophical psychology can expect to gain any deeper positive understanding of the psychology of human beings without some empirical input. A Rylean de-construction of the ghost in the machine is good as far as it goes, but it does not, for instance give us any positive insights into the phenomenon of human consciousness.

Another possible claim could be: "It is true that some empirically based knowledge is necessary for bioethical arguments, but this knowledge does not have to be based on empirical research. Anecdotes, asking around the circle of acquaintances and introspection is a sufficient basis." Again it is true that not all the empirical knowledge needed for bioethical arguments need to be based on empirical research, but almost certainly false that all the necessary knowledge can be "research independent".

Different Uses of Empirical Data

Until now we have just been talking about "the use of empirical data" but there are clearly a number of different uses and a number of different types of data, and some uses are much more contentious than others.

In the following I will discuss both the contentious and the less contentious uses.

Empirical data and moral theory

The most contentious use of empirical data would be to use them either to place restrictions on moral theory, or to provide positive resources for moral theorising.

The clearest exponent of the use of empirical data as a source for restrictions on moral theory is Owen Flanagan who in his "Varieties of Moral Personality" argues that psychology does place limits on the kind of moral theories that can be projected as valid moral theories for human beings. His main argument is that certain moral theories require moral agents to have psychologies that are unrealisable by human beings. Direct act consequentialism requires computational powers that we do not have, and cannot ever get; and certain kinds of virtue theory require unrealisable combinations of virtues. Flanagan does not argue that psychology uniquely determines the one and only correct moral theory, but he does argue that psychology delimits an area of acceptable theories.

The most recent exponent of the positive use of psychology is Jonathan Glover in his book "Humanity – A Moral History of the 20th Century". Here Glover argues that human psychology contains certain moral resources, morally positive psychological traits and features, and that it is one of the tasks of moral philosophising to take account of these resources, and make sure that 1) our moral theories make full use of them, and 2) that our social systems promote them.

Both Flanagan and Glover are fairly dismissive towards the moral philosophers who think that moral philosophy can be done from the armchair without real engagement with empirical psychological reality.

It is impossible here to go into the finer detail of the arguments presented by these two philosophers, and unfortunately also impossible to look at the important differences between them. Here we will have to move straight to the possible counter argument against both the Flanagan and the Glover project. This counterargument goes something like the following: "All this is very well, but it is of little relevance to moral philosophy. Moral philosophy is primarily prescriptive. Our moral theories should tell us what to do, and why, and it is basically irrelevant whether or not our psychologies make it hard or easy for us to do the right thing."

Can this challenge be met? This depends on whether or not the exponent of this challenge holds a version of the "ought implies can" principle. For the hard-headed opponent who is willing to claim that "ought" and "can" are totally unrelated, it is clearly possible also to maintain that "ought" and "psychological realisability" are also unrelated. However, for the opponent who accepts that in certain circumstances the fact that I cannot perform an act entails that I have no moral obligation to perform it is left in the more difficult position to explain which kinds of "cannot" implies which kinds of "no obligation". If some physical limitations on human agency are sufficient to weaken the force of some moral claims, why is the same not true of some psychological limitations?

Influencing the problem space of bioethical considerations

In this section I want to consider a potentially less controversial use of empirical data, and we will turn from psychology to sociology. A scanning of the bioethical literature will show that bioethicists work on a very diverse range of problems. Some of these are problems that actually do occur in the relevant field of practice, some are problems that have not yet occurred, and some are problems that will probably never occur. It is also obvious from such a scanning that there is no simple relation between the importance of a problem in practice and the amount of bioethical work being done on the problem. Some problems seem to grasp the attention of bioethicists, not because they are practically important, or because they affect large numbers of people (or in some cases any people at all) but simply because they nicely exemplify certain theoretical points, or are useful for sharpening some theoretically interesting distinctions.

I take it as uncontroversial that at least part of the *raison d'etre* of bioethics is to make a positive influence on what happens in the world, how people act, and what outcomes these acts have. We do, for instance, not primarily teach bioethics to health care professionals in order to make them into philosophers, but in order to make them better health care professionals, who treat patients in an ethically more appropriate way. We can therefore raise the question of what would make an ethical problem an important ethical problem seen from this use-perspective of bioethics. There are a number of possible answers. A problem can become a practically important problem if it:
1. Is frequent
2. Has serious consequences for some people
3. Bother practitioners (e.g. by being a "hard problem")

It is fairly obvious that 1. & 3. are primarily accessible through some kind of empirical investigation of the frequency and perceived hardness of different types of problems; and in some cases 2. also has to be empirically substantiated.

A bioethicist who really wanted to make an impact, to make the world a better place to live, would presumably want to work on the practically important problems, and not on those which only have theoretical interest. This consideration should be especially important for consequentialist bioethicists, whose moral theory directly instructs them to pursue the line of work that will maximise good consequences.

A possible counterargument would be that in bioethics we can also employ a distinction similar to the distinction in many sciences between basic and applied research, where the theoretical work would be the parallel to basic research and the work on practical problems the parallel to applied research. This observation is probably correct, but immediately raises 1) the problem of the balance between the two kinds of research, and 2) the question whether moral philosophy in general should not be seen as the basic research of bioethics.

Making bioethical analysis realistic

A further possible use of empirical data in bioethical argument is to make such argument, and the conclusions reached realistic. It is a feature of lot of bioethical arguments that the argument does not only reach a conclusion of the type "Acts of type X are ethically unproblematic" (a pure ethical conclusion), but that the author goes further and suggests how acts of type X should be regulated, or how the health care system or other sectors should be organised in order to promote such acts. The question of which form of legal or quasi-legal regulation best promotes or supports a certain desired goal is not a purely conceptual question, and the question of which form of organisational re-engineering can create desired changes is even less of a purely conceptual nature.

At the very banal level, any bioethicist considering the importance and practice of informed consent in the A&E department on a practical level will have to take account of the time constraints and other pressures experienced by A&E staff. Otherwise the analysis will become so divorced from reality, that it will have no possibility of making a practical impact, and it could conceivably also add to the (surely totally misguided) view that bioethicists are ivory tower academics with no understanding of real life. Similarly, any bioethicist wanting to promote ethically appropriate whistle-blowing in the health care system will do well to consider the effects of professional hierarchies on junior members of these hierarchies.

Good regulation and good organisational design requires both psychological and sociological knowledge which is more substantial than mere anecdote or personal experience. It also requires new empirical research to check whether the system that has been put in place actually works, since the history of regulation and organisational re-design is filled with examples of well-intentioned attempts that simply do not fulfil their purpose. For the bioethicist the important consideration must be the eventual outcome, not in terms of regulations or organisations, but in terms of what actually happens in the world. But it is crucially important that such outcome research is bioethically informed, and asks the right (i.e. morally important) questions.

Some Final Comments About Consequentialist Doublespeak

Some of the most outspoken critics of the use of empirical knowledge in bioethics belong to the broad school of consequentialists. This is somewhat surprising since most modern consequentialists hold some version of preference consequentialism, where the consequences to be maximised are spelled out in terms of preference satisfaction. There are many variants of this theory with different requirements for how well considered preferences have to be to count (do they for instance have to be rational – whatever that means), or whether or not other regarding preferences count etc. etc. and there are also a number of variants with regard to whether consequentialism should be maximising, satisficing or some even more complex variant. Now, preferences, their distribution, their justification, their strength, and their dependence on various social influences seem to be exactly the type of questions that are open to empirical research. And if we were really in the business of maximising preference satisfaction we would need the data generated by such research. But consequentialists usually do not go down that road. For them numbers only count in stylised examples, or when we are on the absolutely safe ground of comparing the miserable lives of millions in the third world with the good lives of citizens in the affluent West.

If we for instance look at the question of reproductive choice we often find arguments by consequentialists that go something like the following. People should have the right to choose how they want to reproduce, because 1) it is a very important decision for them, and 2) denying them that right amounts to unjustified discrimination.

The first of these considerations is a *bona fide* consequentialist consideration, but could conceivably be outweighed if a sufficient number of other people had weaker preferences against this specific form of reproduction, whereas the second consideration is not truly consequentialist at all. Unless it can be shown that holding preferences about how other people reproduce is irrational, it is hard to see that a consequentialist claim of discrimination can be justified. Most consequentialists believe that everyone should count as one and no one as more than one, but this may in some cases lead to a majority imposing its will on a minority.

It is not sufficient to show that it is impossible to justify such preferences rationally. Many of our important preferences have no further justification than "because it gives me pleasure" or "I find it abhorrent". For most people their choice of food is, for instance not governed by an elaborate nutritional calculus, but by "mere taste and preference". The consequentialist would have to show that the preferences in question were positively irrational, and this amounts to a very difficult task. Individuals may have bad justifications for such preferences, and many undoubtedly have, but for others it may simply be a preference which even after sustained scrutiny and consideration seems to be an acceptable part of their preference structure. The immediate retort that ethics cannot be built on gut reactions but only on reasoned argument is absolutely right but misses the point. The structure of the preferences we are discussing is not "I don't like X, therefore X is wrong", but "I don't like X, so allowing people to do X reduces my wellbeing". In the conseqentialist case we have a theory which has been developed and justified by reasoned argument, and this theory takes as one of its input parameters the preferences of people (not the preferences of philosophers). One of the great attractions of consequentialism is that it works with people's own definitions of their wellbeing and does not seek to impose a unified picture of the good life at the theoretical level. But in many actual avowedly consequentialist arguments, this commitment to work with people's own preference structure is forgotten, and non-consequentialist considerations are brought in to remove undesired results.

This could lead one to suspect, that consequentialist philosophers seek to reach laudable liberal conclusions, by refusing to take their own theory and its analysis of preferences seriously.

Health Care and the Habermasian Public Sphere

Dr. Andrew EDGAR
*Centre for Applied Ethics,
University of Wales, Cardiff*

The purpose of this paper is to explore the empirical pre-conditions that must exist for publicly accountable health care system to be possible. This will be done by rehearsing Habermas's arguments concerning the structural transformation of the 'public sphere'. The paper presupposes that the public accountability of health services is politically desirable, and thus that the public should be involved in debates over the organisation of health care and access to health services. However, Habermas's scepticism over the fate of rational and critical public debate in late capitalist societies will be used to question the degree to which the political ideal of a publicly accountable health service is achievable today.

I

The 1980s and '90s saw both an increasing impetus towards the reform of the Western health services, and a parallel growth in interest in the possibility of public consultation over that reform. While the gauging of public opinion about health care policy is not new - for example, the Beveridge Report of 1942 may be seen to respond, at least in part, to public opinion surveys that had been carried out by and on behalf of the British government before and during the war [1]- a renewed appeal to public opinion in the 1980s may be understood on the one hand, in terms of the need to legitimate the fundamental reform of health care systems in the face of a crisis brought on by increased patient expectations, an ageing population, and inflation in the costs of providing health care. On the other hand, public consultation may be understood as part of a reaction to the perceived paternalism of the medical profession, and thus as part of a broader movement to increase the say that patients have in their treatment. Thus, while large scale public consultation exercises, such as those carried out in Oregon, New Zealand and the Netherlands, may have played a part in guiding and legitimating policy change, other phenomena, such as a renewed interest in medical ethics, the propagation of patients' rights, and the quality of life movement, sought to shift thinking about medical treatment and medical outcomes away from the typically physiological criteria that were the preserve of the medical professions, to criteria that took account of the patient's experience of illness and treatment.

Approaches to public consultation were diverse, ranging from postal and telephone surveys of large sample populations, through interviews, to community meetings, focus groups and citizens' juries. The questions and problems posed to respondents were equally varied, from the consumerist issues of patient satisfaction, through policy issues concerning the prioritisation of treatments, of conditions and of types of patient, to deeper moral and political concerns with the eliciting of the values that underpinned public attitudes to the health service and its policy. In parallel, much research into health-related quality of life was concerned with the way in which patients and lay people understood health and the impact that illness had upon their lives, as well as with the more overt issue of the relative weights that they might give to diverse health states [2]. If there is variety in the approaches and questions posed, then there is also variety in the policy issues that are at stake in consultation exercises. In Oregon, public consultation was carried out with the explicit intention of excluding certain services from Medicaid coverage, and thus was concerned with rationing in the strictest sense of the word. In contrast, the UKNHS refused formally to relinquish a historical commitment to universal access to all forms of health care. The diverse consultation exercises carried out by UK Health Authorities had therefore to struggle with a much more diffuse and ill-defined notion of prioritisation, which effectively entailed identifying certain treatments or conditions as more or less important than others, without the possibility of asserting explicitly that some treatments should be denied public funding altogether.

While there exists a large and diverse body of practical and theoretical work on public consultation, it may still be suggested that a number of crucial social and political assumptions have not yet been adequately articulated. The development of public consultation has occurred within the context of an often deep ambiguity between the status of the patient as citizen and as consumer. In the UK, this occurs not least through the publication of the white paper *Patients First* in 1979, and the introduction of internal markets in the NHS. Similarly, while a number of studies (and not least that of the Core Services Commission in New Zealand) have been sensitive to the pluralist nature of contemporary societies, the implications that this pluralism has for articulating a coherent concept of 'community values' have still to be worked through. To make this point is, in effect, to ask exactly what it is that is being appealed to in the name of public opinion, and whether it is possible - or indeed desirable - to harmonise the different values that a consultation process will elicit. Finally, Ann Bowling et. al. have posed perhaps the most disturbing question of all: What is to be done if, having consulted the public, one does not agree with the answers it gives? [3]. The appeal to Habermas's account of the development and decline of a bourgeois public sphere may begin to clarify these issues.

I

Habermas defines the public sphere as the coming together of citizens to express their opinions about matters of general interest. The communication can be face to face, or can be through letters, journals and newspapers or the electronic mass media. Crucially, the mere expression of subjective opinion is insufficient to constitute a public sphere, or to give rise to genuinely public opinion. Opinions must be subject to rational debate, and a citizen's participation in the public sphere depends, in no small degree, upon their willingness to have their own opinions subject to this rational scrutiny, and on their part, to scrutinise the opinions of others[4].

What Habermas terms the 'bourgeois public sphere' emerges in the seventeenth and eighteenth centuries in Britain, France and Germany. Its origin lies in the emergence of a newly affluent, well-educated and leisured professional middle-class. Habermas suggests that it is this class's relationship to art and literature, and particularly to new literature, that central to the formation of a public sphere. A literate and leisured class has the time a resources to consume art. Art thus shifts from being a concern of the state and church - as means to the ostentatious public display of power -to something that is increasingly enjoy in private. A market for art emerges, displacing the patronage of the feudal aristocracy. T most archetypal art form of this period is the novel, emerging in England with the work of I Foe, Swift, Fielding and above all, for Habermas, Richardson. That this art is new significant. Previously the literate would, Habermas suggests, have predominantly read t classics. Their status as worthwhile art was not in doubt. New art does not come with th same guarantee. Its status thus becomes something that is open to dispute. Indeed, Fielding *Shamela,* written in reply to Richardson, indicates how dispute was inherent to the growth the novel as an art form [5].

The mere assertion that one derives subjective enjoyment from a work of art insufficient to justify one's choice. The conventions of polite society required that the read is able to discuss and justify their taste through rational argument. This is also, consequence, the period in which the professional art critic emerges, whose task is not mere to identify faults in art works, but also, and more importantly, to educate the public in go taste. Hume refers, significantly, to the 'embarrassing' problem of choosing an appropria critic for one's guide [6].

The new middle-class is not, however, merely a reading public. Its literacy is express in writing, and the pre-eminent form of personal writing is the letter. Habermas suggests th Richardson's *Pamela* is significant not least because of its epistolary form [7]. It is original offered, not as a story, but as a set of exemplary letters. In the letter, the private bourgeo comes to articulate their innermost thoughts and feelings, and to present them in publi Habermas notes that the letter was not, in the eighteenth century, treated as a priva document. Letters would be read by others than their addressee, and indeed would I circulated and even published. The letter thus exemplifies the complexity of bourgeo subjectivity. It has, on the one hand, a new privacy. An intimate inner life can be articulate and the novel is central in providing characters who may be models to assist in th articulation. Yet, on the other hand, that articulated intimate life is made public, and is place under public scrutiny - just as, in effect one's seemingly subjective literary tastes are subje to rational and public debate. Thus, Habermas suggests that the literary public sphere serv to form a distinctive subjectivity that crucially possesses the ability rationally to scrutini itself and the opinions and actions of others. It is thus a subjectivity that has been honed f participation in political debate.

The political public sphere may be seen to emerge as journals, that were originally source of reflection on art and morality, come increasingly to deal with politics (in, f example, the Tory *Craftsman* and *Gentleman's Magazine*). In addition, the public sphe becomes institutionalised in coffee houses and salons, where political ideas and opinions c be debated, alongside the exchange of commercial information or the exercise of good tas in the arts. Parallel developments can be seen in political philosophy - as from Locke throug to Burke there is a gradual overcoming of the suspicion of public opinion, and thus a increasing recognition that those who are governed should have some say in that governmer The press is gradually freed from state control, and access to parliament is increasing allowed to journalists. In Britain, by the beginning of the nineteenth century, there is a increasing recognition that government should act in public, and that it should be responsible to public opinion, not least where that opinion differed from what was formally expressed in voting behaviour [8]. A political public sphere, embodying an ideal of open debate, is the context within which that public opinion is formed, and thus the context within which government policy is subject to scrutiny.

Habermas is, however, at pains to explicate the deep ideological rift that underpins the bourgeois public sphere [9]. On the one hand, the bourgeoisie offer an aspiration to justice, as the public sphere treats all as equals, and respects a contributor purely on the grounds of the rationality of their contribution, and not upon any extrinsic aspects of their social or personal existence. This is, in effect, to offer an image of citizenship. On the other hand, this image conceals the actual processes of exclusion through which the public sphere is constituted. The ideological self-understanding of the bourgeois subject rests heavily upon a notion of its autonomy. It understands autonomy as a universal property of human nature. It thus conceals from itself the material and social conditions of its autonomy. Crudely, it projects the particular conditions of the bourgeoisie as the conditions of humanity as such - the citizen is thus bourgeois. Habermas suggests that bourgeois autonomy in fact rests upon the class's economic position, as the owners of productive property, rather than as labourers. In effect, free and rational debate is made possible, precisely because of the economic homogeneity of the bourgeoisie. Habermas implies that major conflicts of interest, specifically to do with the economy, are excluded from debate. These are resolved through the market, and not through politics. But this entails that the fundamental class conflict between bourgeoisie and proletariat cannot itself be a topic for the public sphere. The two classes lack sufficient common reference points to engage in debate. The claims of the proletariat can thus be excluded by the bourgeoisie as simply irrational. Further, by assuming that the free market, and more specifically, a condition of perfect competition between small producers, is a natural condition of human society, the bourgeoisie can argue that all human beings have the opportunity to enter into the property owning and entrepreneurial class. Thus, once again, the role that the public sphere plays in promoting the narrow class interests of the bourgeoisie, as opposed to the universal interests of humanity as such, is concealed.

The structural transformation of the public sphere occurs with the transition from nineteenth century capitalism, characterised as it is by relatively small scale production, to the increased capital concentration and rising importance of finance capital in the twentieth century [10]. In addition, while the bourgeoisie had largely confined the state to a 'nightwatchman', leaving the organisation of society to the supposedly natural conditions of the free market, twentieth century capitalism may be increasingly characterised in terms of an interventionist state, and thus the development of state welfarism. Eighteenth century distinctions between state and civil society, and thus between the public sphere of politics and the private sphere of the economy and the home are undermined.

On the one hand, economic conflicts come to play an increasing role in politics, as the franchise is extended and mass involvement in politics increases. But, given his assumption that economic conflict lies too deep to allow of rational resolution, Habermas suggests that the expanded public sphere must change character. Rational debate is increasingly replaced by a process of haggling and bargaining, that itself reflects economic exchange. On the other hand, and perhaps more fundamentally, the subjectivity that the bourgeoisie constituted for itself is also threatened. Significantly, the material base of bourgeois autonomy is undermined. The autonomous small-scale entrepreneur is replaced by large joint stock companies, and the bourgeoisie increasingly become functionaries of these companies. The proletariat, equally, come to see themselves in relation to large commercial and state bureaucracies, with both coming to intervene in more areas of their previously 'private' life. The intimacy of the bourgeoisie is thus lost.

While a newly affluent and secure proletariat has more time and resources to enjoy it leisure, these new material conditions, Habermas argues, brings about a destructive transformation in the literary public sphere [11]. While bourgeois taste was grounded in debate, Habermas suggests that modern consumption of culture rests overwhelmingly upon the mere exercise of personal preference. To judge a work good, is to say no more than that one has a subjective liking of it. In effect, Habermas is suggesting that in societies that place ever increasing emphasis upon instrumental rationality, and thus upon the calculation of the most efficient means to achieve any given end, the intellectual resources necessary to debate the desirability of ends are lost. The argument has a close parallel in MacIntyre's criticism of emotivism, and indeed both Habermas and MacIntyre are suspicious of the continuing influence of positivism's marginalisation of the meaningful discussion of values, albeit that Habermas more explicitly ties the emergence and influence of positivism to the development of modern capitalism and its attendant bureaucracy and technocratically orientated science [12]. Thus for Habermas, if the subject is not being schooled in rational debate and self criticism in its consumption of art and culture, then it is not being made ready for participation in political public debate. In effect, if modern subjects respond to culture a mere consumers, then that is also how they will respond to politics. The competencies demanded by citizenship are thus eroded.

Contemporary politics are characterised for Habermas, not by rational debate, but by the political use of advertising, specifically in the form of public relations [13]. He claims that public relations cannot be used simply to manipulate an electorate. The subjects of contemporary capitalism, for all their weaknesses, are not yet reduced to mere dupes of the system. Subjects can and do express their needs, and if they perceive that a political administration is failing to meet those needs, then they will vote against it. Public relations is therefore a means by which policy makers can present their work to the public. Public relations thus maintains a continuity with the demands for publicity established by the bourgeoisie in the late eighteenth century. However, the eighteenth century idea presupposed that the public was addressed as a whole. The very possibility of debate served to form public opinion, and thus to facilitate a collective response to policy. In contrast, and in line with commercial advertising, public relations addresses individual consumers. Rational debate and reflection is not required, only an aggregate of individual expressions of preference. Further, Habermas suggests that public relations works precisely by concealing the partial or private interests that underpin the policy formation, and thus to stifle counter arguments or counter claims through a more or less spurious appeal to collective welfare, and the seeming objectivity of the facts that are evoked in support of the case. In effect, spurious homogeneity is created in the public, mirroring that of the bourgeois public sphere and serving to exclude oppositional claims as irrational, not least because they fall outside the terms of reference created by the public relations exercise.

In summary, Habermas offers a grim account of the decline of the public sphere, from an ideal of rational debate by citizens, to the mere expression of subjective preference by isolated consumers. He is arguing that the conditions necessary for genuine democracy, which is to say, for full public participation in policy formation, are absent in contemporary society. While his later work has pulled back some way from this uncompromising and excessively negative account, its broad parameters may still allow a critical framework within which to comment on the public accountability of health care services.

II

With the exception of surveys of patient satisfaction, it may be suggested that public consultation over health care typically assumes that the public should be treated as citizens, and not merely as the consumers of health care. While the concept of 'citizenship' remains contestable, it may be suggested that, at least in the context of a welfare state, it entails a complex relationship of mutual obligation between the subject and the state. The state has the obligation to protect the citizen, and thus, all other things being equal, to provide health care as it is needed. Yet, as Habermas has suggested of the welfare state, state provision also entails state intervention in the private life of the citizen. Health policy may place certain restraints upon individual freedom, for example in pregnancy, child birth and child care. While these restraints may overtly be presented as being in the subject's long-term interests, Habermas's view of democracy - which at the time of *Structural Transformation* was heavily indebted to the radical constitutional theorist Wolfgang Arbendroth - implies that such intervention requires citizens to have some say in policy formation. An ideal, whereby those who are governed formulate policy, is aspired to. Yet, beyond the relationship of the citizen to the state, there remains a further relationship of citizens to each other. It may be suggested that if a health service is understood politically, which is to say, in terms of a provision to citizens and not merely to consumers, then citizens have an obligation to each other to ensure that their personal demands upon the service are not excessive. It is plausible that the UK's reliance upon waiting lists can be understood in this way. A waiting list is acceptable, even to those upon it, providing those higher on the list have some greater demand upon the scarce resources of the service. To wait one's fair turn, however that fairness is to be understood, is expressive of a civil responsibility.

The implication of *Structural Transformation* must be that appeals to citizenship are now largely illusory. The shift to consumerism in healthcare, for example with *Patients First*; the institution of an internal health service market, and the publication of a document of patients' rights in the UK, is not a product of government policy as such, but rather government acknowledgement of the true state of affairs. Government espousal of consumerism therefore has a grim honesty about it, and entails a refusal to allow a spurious concept of citizenship to taint the public relations presentation of policy. From this perspective it may be suggested that exercises such as the Oregon experiment and the work of the New Zealand Core Services Commission are little more than public relations exercises, that appeal to illusory community values and illusory notions of collective welfare. In a similar vein, Toth indicates that the institution of the citizen (as distinct to the consumer) as the primary recipient of health care, may militate against public control of health care. He suggests that the emergence of the NHS in the UK worked in large part to free the medical profession from interference by patients (rather than to increase the control of patients and the public over medicine), and that the NHS has done little to improve the health of the general population. The patient as consumer, prior to the Second World War, had a significant degree of influence over the doctor. Once medicine is provided from within the welfare state, the medical professions free themselves from any accountability to lay bodies [14].

Yet it is precisely the reasons why appeals to citizenship might be illusory that are significant. Habermas's argument in *Structural Transformation* was that the intellectual resources that subjects require in order to debate and reflect critically are denied to them. While Habermas's critical comment that modern elections are little more than plebiscites on government policy may be accepted, this may still be understood as an institutional failing. Elections are crude, and perhaps increasingly irrelevant, mechanisms for eliciting public opinion. Public opinion may, none the less, be formed through debate - and in the UK repeated calls for a debate about health care (and indeed many other areas of policy) seem to echo this belief. Increasingly, it may be suggested, health care is something upon which subjects have reflected and debated. Such debate is mediated by news coverage and the comment columns of newspapers, but also by fictional portrayals of the health service on television. However, the degree to which such mediation leads, not to open and rational debate, but to what Habermas subsequently referred to as systemic distortion, remains an open question. Habermas's profound scepticism over the processes of public relations, and indeed the implication of journalism in advertising, remains pertinent. One may ask, for example, to what degree the fragmentation of the audience for newspapers and the electronic media in terms of patterns of consumption (thus, in effect, market segmentation), at once influences the form and content of the opinions offered for debate to each fragment, and crucially, the degree to which such fragmentation reflects deeper, unarticulated, conflicts of interest in the society.

Habermas's comments upon the limits of rational debate, and thus the formation of a public sphere, may be of relevance. Firstly, he suggests that the public sphere presupposes a certain homogeneity in its participants. The Oregon experiment presupposed such homogeneity, seemingly assuming that there was something characteristic about Oregonians, and that distinctive communal values could be elicited. While this may be true, even here a general problem of Oregon's approach to the eliciting of communal values was that community meetings tended to attract an overwhelmingly professional audience. The proletariat, and thus the principal beneficiaries of Medicaid, were excluded, or excluded themselves. The problem of pluralism is acute in the UK, where the jurisdiction of local Health Authorities need have little or no correspondence to a culturally homogeneous community. Appeals to community values appear to be largely spurious, if this entails neglecting the values of marginal or minority groups within the community. New Zealand's approach, of targeting different social groups is more interesting, but then leaves the question of how diverse values, and perhaps more significantly, conflicting grudges and discontents are to be managed and resolved. How does one respond to a Maori's simple solution to the problem of scarce resources: reduce the health state of the whites to that of the Maori? [15] This is significant not least because it suggests that, underpinning debate about health care there is an economic struggle over resources. Indeed, Habermas notes that, within a welfare state, health care may be the sort of issue that serves to undermine the legitimacy of capitalism when more traditional areas of political conflict have been successfully managed by the state [16].

Reflection on the limits of rational debate may be taken further, to ask if health care is itself something that can be rationally debated. Habermas's model of the bourgeois public sphere presupposes the economic autonomy of the bourgeoisie. The proletariat is excluded from the public sphere, precisely because of their lack of autonomy, and more problematically, this lack of autonomy is something that cannot be subject to rational debate. Yet, in the experience of illness, it is precisely personal autonomy that is at stake. Illness compromises one's competence as a social agent. At the core of medical ethics there lies a recognition of the need to protect the vulnerable patient from the autonomous power of the medical profession. This begins to raise questions about the sort of resources that are then needed to pursue rational debate in such areas of powerlessness. The experience and threat of morbidity and mortality may be understood as phenomena that fall outside rational debate, precisely because, as with the experience of the proletariat, sound health is a precondition of rational autonomy. Something of this is expressed in John Harris's criticism of rationing according to QALYs or years of life remaining to the patient. In a choice over access to life saving health care all contenders are threatened with the same loss: the rest of their lives [17].

Habermas, in the critical reflections that he published on *Structural Transformation* [18], points to the role that forms of identity politics can play at this point. Identity politics, which can find ways to articulate the experience of the oppressed and marginalised, serves to shift the parameters of rational debate, and to expose the power structures that underpin received opinion as to the nature of rationality. In terms of health, this would entail a renewing of a culture's capacity to talk of mortality and morbidity. It would, in Habermas's own terms of communicative competence, demand not merely that the healthy learn to listen to the sick, but also that the sick learn to listen to each other. Only thus would there arise an articulate, but also critical discussion of the experience and implications of illness in its many different forms.

Habermas notes that *Structural Transformation* focuses upon the bourgeois public sphere at the expense of what he terms the plebeian public sphere. He is initially sceptical that a proletariat, bound as it is by the constraints of labour, has either the resources or the leisure time to indulge in critical debate. This has been countered, not least by Negt and Kluge's substantial study of the proletarian public sphere [19]. If the analogies between lack of political autonomy and lack of physical or mental autonomy drawn above are worthwhile, then it may well be that Negt and Kluge now have more to offer in understanding the nature of rational debate over health care than has Habermas.

IV

Three problems were posed at the end of section I above: the tension of citizenship and consumerism; the question of the nature and reality of public opinion; and the question of what is to be done with problematic results of consultation exercises. The first two issues are, it has been suggested, intimately linked. Public opinion relies upon citizenship, at least insofar as the citizen is understood as a rational and critical disputant. A politically conceived health care service presupposes that health care policies are subject to critical debate by citizens. Yet, such debate can be distorted, not merely by the illusions of public relations and deeper political and economic conflicts, but also by our very inability to speak coherently and articulately of the experience of illness and death. Yet it is only through the development of theory that can explain and facilitate the recognition of the sources of distortion in debate that a response can be provided to the third question. For the policy maker to find themselves in disagreement with public opinion demands that they ask, on the one hand, if the results of the research are genuinely expressive of public opinion (and not the non-public opinion of consumers, that is elicited by market research), and on the other, how, if it is public opinion it poses a rational criticism of the opinions and values of the policy maker. Yet neither the genuineness of public opinion nor the rationality of its critical intent will be self-evident from the mere empirical data provided by social surveys or other forms of public consultation. This alone serves to emphasise the need for theoretical reflection - both upon the nature of the welfare state and a national health service, and upon the nature of political debate. It also emphasises the vulnerability of expressions of public opinion to being neutralised by that very theory. Further reflections on the work of Habermas, and collaborators such as Negt and Kluge, and Apel and Offe, may serve to refine theory into a tool that can serve for the just interpretation of public opinion data, and thus contribute to the realisation of a public sphere relevant to health care.

References

[1] Toth, (1996), p.181
[2] Brooks, (1995)
[3] Bowling, (1993)
[4] Habermas, Jürgen, (1974), pp.49-55
[5] Habermas, Jürgen, (1991), pp.43-51
[6] Hume, David, (1985), p.241
[7] Habermas, (1991), pp.57-67
[8] Habermas, (1991), p.49
[9] Habermas, (1991), pp.79-88
[10] Habermas, (1991), pp.141-151
[11] Habermas, (1991), pp.159-75
[12] MacIntyre, (1981); Habermas, (1971)
[13] Habermas, (1991), pp.211-222
[14] Toth, (1995), pp.184-5
[15] Campbell, (1994), p.11
[16] Habermas, (1971), p.108; Habermas, (1991), p.157
[17] Harris, (1987)
[18] Habermas, (1992)
[19] Negt and Kluge, (1996)

Bibliography

Brooks, Richard G, (1995), *Health Status Measurement: A Perspective on Change*, London, Macmillan
Bowling, Ann, Jacobson, Bobbie & Southgate, Leslie, (1993), 'Health Service Priorities: Explorations in Consultation of the Public and Health Professionals on Priority Setting in an Inner London Health District' *Social Science and Medicine*, 37(7), pp.851-857
Campbell, Alastair, (1994), *Ethics Workshops: Public Participation in Discussing Ethical Issues in Defining Core Services*, Wellington, Core Services Commission

Habermas, Jürgen, (1971), *Toward a Rational Society*, London, Heinemann
Habermas, Jürgen, (1974), 'The Public Sphere', *New German Critique*, 3
Habermas, Jürgen, (1991), *The Structural Transformation of the Public Sphere: An Inquiry into a Category of Bourgeois Society*, Cambridge MA, MIT Press
Habermas, Jürgen, (1992), 'Further Reflections on the Public Sphere', in Craig Calhoun (Ed.), *Habermas and the Public Sphere*, Cambridge MA, MIT Press
Harris, John, (1987), 'QALYfying the Value of Life', *Journal of Medical Ethics* 13, pp.117-23
Hume, David, (1985), 'Of the Standard of Taste' in his *Essays, Moral, Political and Literary*, (ed. Eugene F. Miller), Indianapolis, Liberty Fund
MacIntyre, Alasdair, (1981), *After Virtue: A Study in Moral Theory*, London, Duckworth
Negt, Oakar & Kluge, Alexander, (1996), *Public Sphere and Experience: Toward an Analysis of the Bourgeois and Proletarian Public Sphere*, Minneapolis, University of Minnesota Press
Offe, Claus, (1984), *Contradictions of the Welfare State*, London, Hutchinson
Toth, Ben, (1996), 'Public Participation: An Historical Perspective', in Joanna Coast, Jenny Donovan & Stephen Frankel (Eds.), *Priority Setting: The Health Care Debate*, Chichester, John Wiley

Putting Empirical Studies in their Place

John HARRIS
Centre for Social Ethics and Policy
University of Manchester

'I cannot forbear adding to these reasonings an observation, which may, perhaps, be found of some importance. In every system of morality which I have hitherto met with, I have always remarked, that the author proceeds for some time in the ordinary way of reasoning, and establishes the being of a God, or makes observations concerning human affairs; when of a sudden I am surprised to find that instead of the usual copulations of propositions, is, and is not, I meet with no proposition that is not connected with an ought or an ought not. This change is imperceptible; but is, however, of the last consequence.'[1]

David Hume's famous articulation of the naturalistic fallacy may not have identified a genuine fallacy since, as many subsequent discussions have demonstrated, values are inevitably enshrined in the language with which we describe and present facts [2]. However, Hume does draw our attention to the necessity of establishing moral conclusions by some process other than the simple accumulation and recitation of facts, followed by a completely unprecedented and non-sequential[i] inference. Hume of course went on to conclude (on no basis whatsoever) that since our "decisions concerning moral rectitude and depravity are evidently perceptions;...Morality therefore is more properly felt than judged of" [3]. Bertrand Russell describes Hume's enterprise as a "self refutation of rationality" [4] and I agree with Russell that Hume's account of morality is unreasonably pessimistic about the possibility of a rational morality.[ii]

The lesson for us today, is not one about the grounding of moral principles, although some lessons will emerge about that subject as well. Rather, Hume offers us a salutary reminder that the relationship between facts and values is by no means straightforward, and that when factual inquiries appear under the guise of ethics, important questions need to be asked and answered.

Bioethics is in danger, not of committing the so-called "naturalistic fallacy", but of something perhaps both more clearly erroneous and more clearly unhelpful. Hume famously accused "all the vulgar systems of morality" of erroneously and unjustifiably deriving an ought from an *is;* but much contemporary bioethics and indeed even more contemporary policy making on ethics, seems to be committing a clearer and more vulgar fallacy, this we might call the "Empiricalist Fallacy".

The Empiricalist Fallacy involves the idea that in ethics one might dispense with *oughts* altogether and so spend time usefully describing or producing an "interminable and inconsistent" series of *"ises"*, "to confound an opponent and glorify oneself"[iii]. Of course facts are fascinating and some facts are even essential, but gathering them is not the business of ethics. Empirical research of any sort is not ethical research. It might be essential to ethics, it may be the result of ethics, but ethics it ain't. Empiricalists believe they can do something which may legitimately be described as ethics without resort to oughts at all, and this belief is simply erroneous.

The Role of Empirical Studies [5]

What then may we say of the role of empirical studies in the domain of ethical research? Clearly science and empirical research is *relevant* to the study of ethics and to ethics research, but how exactly? If we first think about the role of science, it is clearly essential to ethics in important ways, for example, we need to know what embryo research is and does or what stem cell research is and what it might be expected to achieve and at what cost. We cannot begin to address the ethics of such research and its applications without knowing the facts in some detail. And of course if we don't know the facts then we have to find them out before we can start assessing the science and its applications from the perspective of ethics. But gathering the facts and *deciding what we ought to think about them* are two different sorts of activities.

In essence this is not a problem about "methods in ethics", this is a problem about the *funding* of ethical research. The place of empirical studies in ethics would not be an issue if it did not vitally relate to the funding of ethical research. So, when a community decides that it needs to be better informed and prepared about the ethical problems that it faces, it will not advance that objective one iota if it puts its money into embryo or stem cell research. Similarly, it will be no further towards its objective if it tries to find out what people think about embryo or stem cell research. At best this will tell a society what *political* or *social* problems it might face in implementing particular policies on these issues or indeed what problems it might face in funding the scientific research, but it will tell us nothing about what should be funded, or what should be done. If we fund embryo research or stem cell research we are deluded if we think we are putting money into ethics. And if, in funding social or economic research relating to these matters, we try to find out what the public thinks about these issues or what they will cost, again we are deluded if we think we are putting money into ethics.

We need the facts, sure! But when we invest in ethics we must invest in the people and the research that will first identify the ethical issues that are raised by such research, and who can then draw some conclusions about what may or may not be done.

If I may quote from a recent paper [6] on the nature and scope of bioethics:
'Bioethics investigates ethical[iv] issues arising in the life sciences (medicine, health care, genetics, biology, research, etc) by applying the principles of Moral Philosophy to these problems. Medical Ethics and Genethics (ethical issues arising from the discipline of genetics) are subsets of Bioethics.'

Methodology

There is often much confusion about what the methodology of Bioethics is. Bioethics is often characterised as a multidisciplinary mode of inquiry. Health care professionals, life scientists, philosophers, theologians, lawyers, economists, psychologists, sociologists, anthropologists and historians are among those who are typically involved in bioethical inquiry. However, while a wide range of disciplines are actively involved in bioethics,

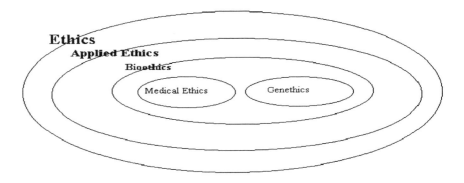

Figure 1: The Relationship between Ethics, Applied Ethics, Bioethics, Medical Ethics and Genethics.

the central method of bioethics is moral philosophical enquiry. Bioethics, rather than being a multidisciplinary mode of inquiry is a branch of applied ethics which is characteristically informed by multidisciplinary expertise and findings. As Ronald Green puts it:

'while ethics and moral philosophy may sometimes represent a relatively small part of the actual work of bioethics, they form in a sense the confluence to which all the larger and smaller tributaries lead, and, more than any other single approach, the methods of ethics and philosophy remain indispensable to this domain of inquiry' [7]

Applied ethics involves the application of the principles and procedures of Moral Philosophy to practical problems. Bioethics, a branch of applied ethics, applies these principles and procedures of Moral Philosophy to issues arising in the life sciences.

Two further issues require some comment. The first is the way in which bioethics is now pursued on the international stage. The second is the precise role of public consultation in ethics. Let's start with what I have come to think of in terms of the globalisation of ethics.

The Globalisation of Ethics

The globalisation of ethics, [8] as I am using the term is the phenomenon whereby the ethical agenda is increasingly set, not by religious, cultural and indeed ethical traditions, nor by competition in the marketplace of ideas, nor by community leaders, exceptional sages or 'saints', nor indeed moral philosophers; but rather this agenda is set by national and international ethics committees, conventions protocols and the like.

Three features of this phenomenon are particularly important.

The first is that the standard of ethical argument where it exists is often poor. Too often, indeed, argument is excluded from the reports altogether. The second feature that should be noticed is that it is these conventions and protocols that are cited in ethical justification, not only of personal conduct but also of national legislation. In other words it is these reports that are fast becoming the reference points for ethical decision-making. This is disturbing because, even where the reports are well argued, the argument is necessarily brief and itself pendant upon other sources. We are in danger of seeing an increasing marginalisation of serious work in Bioethics and an increasing use of, and reliance on, reports and other relatively brief public statements of various sorts.

The final telling feature of the globalisation of ethics concerns the ways in which these international conventions are arrived at.

Classically they are the products of high-level meetings, so they should be of course, but such meetings must achieve consensus, and consensus can often only be marshalled around high-minded, resonant and increasingly abstract principles. Too little attention is paid to the consequences of these principles when applied to particular circumstances, and to the ways in which they may be incongruous with other equally widely held and respected principles that have escaped formal articulation in international conventions.

Very often they also involve so-called "consultation exercises" through which attempts are made to find out what the public feels about the issues or would do about them by way of legislation. This is fine, because it is important to know where we are starting from, so to speak. All too often however, the results of a consultation exercise are taken as powerful indicators of where we ought to end up. And where the consultation exercise has been undertaken by a body reporting to a national government or other elected body, these results cannot help but be taken as indicators of what the voters want and therefore what it would be "good" to deliver by way of any resulting legislation or regulation.

The only point in assembling committees of those whom we tend to call in Britain "the great and the good" is that such great and good people should give a lead on the great issues of the day. All too frequently however, they feel constrained by political prudence and the need to achieve consensus, to follow rather than lead public opinion.

This tendency towards conservatism is often thought to be consistent with prudence and with the so-called "precautionary principle" which, while vague, is generally interpreted as requiring, not unreasonably, that we always proceed responsibly, exposing communities and one another to the least danger. However, the line of "safety" is not always co-extensive with a policy of minimum change. Very often the failure to take radical steps is the policy that costs us most, both in the short and in the long term.

Of course, whether this is true and the extent to which it might be true in any particular case is just what we want committees of the great and the good to assess for us [9]. And having assessed this of course we need such committees to set out the ethical arguments as to whether, in the light of such assessment, it might, firstly, be ethical and secondly, prudent to pursue the research or the policies in question.

What is the Role of Public Consultation in Bioethics?

There are many activities which fall broadly under the heading of "public consultation".
Some are structured meetings or events, for example:
- Focus groups
- Consensus conferences
- Public meetings
- Citizens juries

Others are information gathering or opinion monitoring exercises, which may include:
- Opinion polling
- Questionnaires
- Telephone sampling
- Structured interviews
- Selective mail shots
- Referenda

What is the role and purpose of such activities in bioethics?

Almost all publicly funded ethics research believes that it should include some attempt to discover what people think about the issues under consideration, but the point and purpose of obtaining this information is seldom clear.

There are important differences between a number of different ideas and activities here. Notably between:

- Public consultation
- Public participation
- Public involvement
- Public access
- Public accountability
- Public control

Further, there are a number of ideas or indeed principles, which it is believed some or all of the above serve. They are:

- Transparency, and
- Openness

None of these are clearly defined ideas and while many of them, like motherhood and apple pie, command fairly universal approbation, it is unclear what point they serve in bioethics research.

We need to be clear about the radical differences that obtain between an obligation to consult and a prudent interest in consultation. You might, for example, want to see what people think so that you know what you might be up against if you are going to recommend something different, or, on the other hand, you might find out what people think so that you can be sure not to recommend anything different. Equally, there is a difference between the idea that it is useful to allow people to *contribute* to the decision-making process, and the idea that you ought to allow people to *dictate* the outcomes of that process.

For a committee or commission that is set up to advise the government of a nation state, like the National Bioethics Advisory Commission (NBAC), now "transmogrifiedalized" by George W. Bush into "The President's Council on Bioethics" in the United States, or the Human Genetics Commission (HGC) in the United Kingdom, there is always an acute danger in public consultation or opinion-taking. A government is unlikely to accept advice that is radically at odds with the views of voters. If those views are discovered in advance of the commission forming its own views (and presenting them, with supporting evidence and argument for public consideration,) then there is a danger that public consultation will simply pre-empt the process that the commission was designed to initiate and by-pass the advice that it is its responsibility to give.

Examples

Let's start with an example of a case in which empirical evidence is clearly required for ethical decision-making.

The Human Fertilization and Embryology Act 1990 states, at Clause 13.5.
'A woman shall not be provided with treatment services unless account has been taken of the welfare of the child who may be born as a result of the treatment (including the need of that child for a father), and of any other child who may be affected by the birth.'

This requirement has led many people to object to the offering of assisted reproduction because the particular assistance in question is alleged to be contrary to the interests of the child. Very often there is absolutely no evidence available which would justify such a claim and in spite (perhaps because of) this, people feel free not only to ventilate their prejudices on the issue but also to prevent or frustrate the reproductive choices of others in the name of this prejudice.'

Here it would be important to know whether or not there is any evidence as to whether the use of a particular reproductive technology in particular circumstances or for a particular purpose might result in effects adverse to the interests of the "child who may be born as a result" thereof. However the provision of such evidence goes no way to elucidate or resolve the ethical issues and it is instructive to understand why.

The act claims that the interests of the child are to be taken into account, presumably because they are of some moral importance – so far so good. The first thing to know is what, if anything, can be said about such interests. If there is no empirical evidence then we know that those who claim to interpret these interests are guessing, venting their prejudices or, just possibly, appealing to something so self-evidently obvious as to require neither evidence nor argument.

But even if there is conclusive empirical evidence one way or the other we still need to form a view about the moral relevance of those facts. This, the empirical evidence alone cannot give us.

It is as if it were to be claimed that child bearing were dangerous and should therefore be banned. It would first be important to establish whether and to what extent it was dangerous, but this would hardly resolve the ethical issue as to whether or not people should be permitted or encouraged to have children – or whether having children was either reasonable or ethical given, or perhaps despite, the costs of so doing.

It was reported recently that:

'Experts have told Health Secretary Alan Milburn that the need for children to know their genetic background and true identity ultimately outweighs the rights of privacy for sperm donors. Every year around 1,100 children in the UK are born as a result of donor insemination (DI). It is thought that many of them are never told about their background and grow up believing that their biological parents are those who brought them up.' [10]

It is reported that these "facts", among others, will influence ministers to recommend removing anonymity from sperm donors [11]. However, we know that there are significant non-paternity rates in the United Kingdom and other countries, that is, rates of births where the children of the family are not in fact genetically related to the person they believe to be their father and who usually believes he is their genetic father. Non-paternity rates are quoted with wildly differing values (from less than 1% to more than 30%). A modest, and probably reliable, figure is 2% [12]. However even at a modest rate of 2% non-paternity rates in the United Kingdom account for over 12,000 births registered [13] annually to men who are not in fact the genetic father. Thus, if there is such a thing as a "need for children to know their genetic background and true identity", then on the grounds of numbers alone we should start with normal families. This might imply an obligation for paternity testing in all families! The mischief and disruption this would cause is clearly incalculable. What price then a so-called "need to know one's genetic origin"! I do not believe there is any such thing but if there is, it is doubtful that the arguments which might sustain it are such as to outweigh the rights of privacy of sperm donors, still less the rights to protection of the privacy of family life. Here of course the empirical evidence tells us nothing. It gives us numbers of sperm donors and estimated numbers of non-genetic fathers but all the ethical work remains to be done!

I cannot resist by ending with a quotation from a local author, not of course very local, but local in the sense of coming from a society, like this one, where travellers contemplating a day trip to the United Kingdom for "sexual tourism", and perhaps also to recruit domestic servants, would regard horned helmets as essential cabin luggage and would seldom travel without a plentiful supply of combustible materials. Talking of the utility of sampling

public opinion on any matter of substance, the famous and fictional Norwegian clinician Dr. Peter Stockman, was quoted as saying:
'The majority never has right on its side. ...That is one of these social lies against which an independent, intelligent man must wage war. Who is it that constitute the majority of the population in a country? Is it the clever folk or the stupid?...I propose to raise a revolution against the lie that the majority has the monopoly on the truth.'

The Difference Between Ethics Based Medicine and Evidence Based Medicine

The goals of evidence based medicine and the goals of ethics based medicine may be different.
Evidence based medicine has to do with:
1. Patient outcomes
2. Public health outcomes

Ethics based medicine has to do with protecting the rights and interests of the parties to the medical enterprise, to furthering important principles or social objectives such as justice and fairness and perhaps equality, and these are only contingently related to patient outcomes or public health outcomes.
Ethics based medicine then concerns:
1. Obligations
2. Rights
3. Responsibilities
4. Justice/fairness
5. Equality
6. Harm or benefit to persons
7. Autonomy

Judgements About Ethics and Ethical Judgements

Discovering what the public thinks about issues of ethical significance is not the same as discovering the ethical values of the public. And finding out what the public thinks it is acceptable to do about issues of ethical importance is not the same as finding out what the public finds ethically acceptable. This is because judgements about matters of ethical significance are not necessarily ethical judgements.

This is not the place for any attempt at a complete account (if that were possible) of what makes something an *ethical* judgment. However, some things can and should be said here which bear both on the question of the methodology of bioethical inquiry and on the role of empirical methods in bioethics.

"Morality" is just one of the normative systems which operate within society, albeit the one to which all others are answerable. Other general normative systems include the rules governing religious observance, rules of good manners or etiquette, and, of course, the legal system. Then there are the rules of particular professions, occupations, corporations or clubs that are often rather misleadingly referred to as codes of professional ethics or corporate ethics. All or any of these normative systems may enjoin or forbid things in the name of morality, and the operation of these normative systems may generate ethical dilemmas. For example, although it is always wrong (incorrect) to break the law, doing so is not always morally wrong. The law requires us to drive on the left in the United Kingdom. Other countries that regulate road traffic differently are not, for that reason at least, morally inferior to the United Kingdom. There is nothing unethical about driving on the right, even in the United Kingdom, except in so far as it is dangerous (or possibly unfair) to do so where others are conforming to the law. If it is morally wrong to commit murder it is so not because law forbids it; rather the law forbids it because it is morally wrong[v].

The terms "right" and "wrong" are characteristically used to enforce the rules of all normative systems so that it is always "wrong" in some sense to defy the rules of a normative system. For these reason many people use a term like "wrong" to describe actions or practices forbidden by a normative system that they happen to accept.[vi] But saying that something is morally wrong implies more than the fact that it is forbidden by a normative system that the agent accepts. It implies that it is forbidden by that system because it is morally wrong or that in addition to it being forbidden by that system it is also morally wrong.

So it is with personal judgements that are not made simply in accordance with the rules of a normative system. When we ask someone whether they think this or that to be "right" or "wrong", the answer may reveal their prejudices or personal aversions or likes rather than information about what may be properly called their morality or their moral views or judgements. Ronald Dworkin has assayed an account of at least some the factors that make individual judgements moral judgements and which can defeat an individual's claim to be acting from moral conviction.

Dworkin notes that if someone purports to be acting from moral conviction or because his morality requires it, he must at the very least produce reasons in support of such a claim. Moreover, those reasons must meet minimum standards of evidence and argument; they must not, for example, be based on prejudice, or on alleged facts that are both false and manifestly implausible, or on a personal emotional reaction such as "it makes me sick!" Moreover, such reasons will suppose a general moral principle or theory. Here the claim to be acting from moral conviction may be undermined by the insincerity of the agent – he clearly does not accept the theory presupposed by the stand he has taken, or he acts inconsistently with this theory or principle in other areas of his life.

Where individuals consider and evaluate the moral claims of others, such claims command respect precisely because they are *moral* claims and are thus distinguishable from personal preferences or prejudices. And Dworkin concludes that when considering my moral claims:

'You will want to consider the reasons I can produce to support my belief and whether my other views and behaviour are consistent with the theories these reasons presuppose. You will have, of course, to apply your own understanding, which may differ in detail from mine, of what a prejudice or a rationalization is, for example, and of when one view is inconsistent with another. You and I may end in disagreement over whether my position is a moral one, partly because of such differences in understanding. And partly because one is less likely to recognise these illegitimate grounds in himself than in others.'

Finally, Dworkin insists that:

'We must avoid the sceptical fallacy of passing from these facts to the conclusion that there is no such thing as a prejudice or a rationalization or an inconsistency, or that these terms mean merely that the one who uses them strongly dislikes the positions he describes this way. That would be like arguing that because people have different understanding of what jealousy is, and can in good faith disagree about whether one of them is jealous, there is no such thing as jealousy, and one who says another is jealous merely means he dislikes him very much.' [20]

While this is by no means a complete account of what it is that makes a judgement a moral judgement, and while any such account would be in some respects controversial, it does establish that there is *something* over and above the fact that it is a judgement about an issue of moral significance that makes a judgement a reflection of the morality of an individual. This gives us some reason to be cautious about claims to the effect that forms of public consultation have revealed the moral attitudes or values of the public and that these must be both respected and used to inform public decision-making. If what is informing public decision-making is a collection of recorded prejudices or evidence of a slavish and uncritical adherence to a sectarian normative system, then perhaps the respectability, if not the authenticity, of the voice of the people requires challenge rather than faithful reporting and incorporation into the decision-making process. This challenge cannot come from empirical bioethics but must come from philosophical bioethics with its long tradition of critical and independent analysis. And we now come to the question of the role of empirical studies in bioethics.

Empirical studies, as we noted earlier, have played an increasingly important role in bioethics, partly because of the felt need to act on evidence of public values and attitudes to the questions and dilemmas of bioethics and partly because of the nervousness on the part of bodies funding bioethics research about the apparent absence of a clearly recognised and simply evaluated methodology in research applications.

Endnotes

[i] Or is it "non-sequitorial"?
[ii] Although it is unclear as to whether Russell was including Hume's ideas on ethics in his characterization of the Humean enterprise.
[iii] With apologies to D'Israeli.
[iv] 'Moral' and 'ethical' are used interchangeably in this paper
[v] Among other reasons.
[vi] So that when opinion samplers ask whether people believe something is right or wrong they may be receiving information about what a particular normative system enjoins or forbids and not information about the moral beliefs of those who accept this system.

References

[1] Hume, David, (1966), *A Treatise of Human Nature* Book III.I. The Everyman Library Edition, J.M. Dent and Sons, London, p.177-178
[2] Wittgenstein, Ludwig, (1963), *Philosophical Investigations* Basil Blackwell, Oxford; Searle, John, (1969), *Speech Acts* Cambridge University Press, Cambridge (especially Chapters 6, 7 and 8)
[3] Hume (1966), op cit
[4] Russell, Bertrand, (1961), *History of Western Philosophy,* George Allen & Unwin Ltd, London p.646. Russell memorably characterised the quarrel between Hume and Rousseau as symbolic: "Rousseau was mad but influential, Hume was sane but had no followers."
[5] Here and elsewhere in this paper I draw on previously published thoughts on these matters See my "The Scope and Importance of Bioethics" in John Harris Ed (2001). *Bioethics: Oxford Readings in Philosophy,* Oxford University Press, Oxford, pp.1-22
[6] Bennett, Rebecca; Erin, Charles A; Harris, John and Holm, Søren (2002), "Bioethics" in Bunnin, N. & Tsui-James, E. Eds. *The Blackwell Companion to Philosophy,* [Second Edition] Oxford: Blackwell, pp.499-517
[7] Green, Ronald M (1990), 'Method in Bioethics', *The Journal of Medicine and Philosophy,* Vol.15 No.2, April, p.182
[8] Harris, John, (2000) "Research on human subjects, exploitation and global principles of ethics", in Andrew D.E. Lewis and Michael Freeman. Eds. *Current Legal Issue 3: Law and Medicine,* Oxford University Press
[9] For other examples of this phenomenon and its problematic consequences see Harris, John, (1997) "Goodbye Dolly: The ethics of human cloning" in *The Journal of Medical Ethics,* Vol. 23 No.6. December. 353-360; Harris, John (1998), "Cloning and Human Dignity" in *Cambridge Quarterly of Healthcare Ethics* Spring.. Vol.7. No.2. 163-168; Harris, John (1999), "Genes, Clones and Human Rights" in Justine C. Burley Ed. *The Genetic Revolution and Human Rights: The Amnesty Lectures 1998* Oxford University Press, Oxford.
[10] *Observer* newspaper 26th January 2003 Page 1 and 3.

[11] Ibid.
[12] The most reliable recent published article on the subject is: Macintyre S, Sooman A (1991). *Lancet* vol 338, p 1151, and ensuing correspondence.
[13] Source http://news.bbc.co.uk/1/hi/health/2570503.stm
[14] Dworkin, Ronald, (1977), *Taking Rights Seriously,* Duckworth, London. Chapter 9. This quotation at pages 252-253.

Can There Be Moral Experts?

Louise IRVING
Institute of Medicine, Law and Bioethics
University of Manchester

Dr. Nina HALLOWELL
Public Health Sciences
Medical School
University of Edinburgh

The implementation of new medical technologies (such as genetic screening and pre-implantation genetic diagnosis) within clinical practice raises a range of ethical questions. Questions that require informed answers. Who should we as a society call upon to provide us with answers to some of the most pressing questions of our times?

At the present time, bioethicists appear to have adopted the role of moral expert. Indeed, it would appear that practically all biotechnologial and medical innovations that are presented to society require the blessing, or at least scrutiny, of the bioethicists. However, in a climate of increasing public scepticism concerning the role of expert knowledge, we believe there is an urgent need to examine the basis of the bioethicists' claim to expertise.

So what does it take to be a moral expert in the twenty first century? The answer to this question is dependent upon some form of agreement about what moral expertise would look like - what kinds of knowledge an individual would need in order to have moral expertise. This in turn, hinges upon further questions about what sort of things we might require moral experts to do. Should they provide normative evaluations of behaviour and events both real and hypothetical, advise on policy and regulation or should they try to determine which problems are to be defined as moral problems in the first place?

Who are the current moral experts?

For reasons both historical and practical, moral philosophers are at the forefront of the bioethics revolution. It is generally accepted that, given the complexity of the questions generated by NMT (New Medical Technologies), such as molecular and behavioural genetics and in vitro fertilization, and the controversial nature of clinical bioethics, we need individuals who are trained in methods that employ objective and scrupulous reasoning. Moral philosophers would seem to fit these criteria, they are regarded as people of good faith, and are already interested in questions of the right and the good, of how we might live well and justly, and hence, they have come to predominate in the field of bioethics. However, whilst moral philosophers may appear to have the requisite skills for tackling these issues, they have also been heavily criticised for what many see as their ahistorical, asocial and acultural approach to bioethical problems. In this paper we intend to examine some of these criticisms with the aim of elucidating the types of kinds of knowledge other commentators feel should be applied when bioethicists consider questions that affect us all.

Criticisms of the current *Status Quo*:

There are many criticisms leveled at the pre-eminence of moral philosophers in bioethics. In this paper we will discuss the most common ones, for example:

- Bioethics unquestioningly accepts the supremacy of natural science as the only truly valid form of knowledge.
- The analytic framework used within bioethics decontextualises ethical issues and does not allow for the flexibility of position that complex moral problems require.
- There is a general lack of consensus concerning the aims and objectives of bioethics. Bioethicists have been criticised for having no coherent end goal and no clear idea of whose ends they are serving, or should be, serving.
- The attempt to systematise bioethics has resulted in the reification of autonomy and informed consent as the prime moral goods without adequate reflection on their limitations and proper debate about whom this actually benefits.
- A limited number of problems are subject to bioethical debate/analysis. There is a lack of reflection about why only certain things present themselves as bioethical problems or dilemmas.
- Bioethics as a discipline is exclusive. It is dismissive of other disciplinary approaches to, or public participation in debate about, bioethical problems. It employs 'Gatekeeping' practices in determining who should be a bioethicist, i.e. who has the right sort of expertise to comment upon bioethical issues.

Taking these in turn, we will briefly comment on these claims.

1) Bioethics unquestioningly accepts the supremacy of natural science as the only truly valid form of knowledge.

It can be argued that, in contrast to other disciplines, such as sociology and science and technology studies, bioethics too readily embraces the validity of evidence-based medicine. Perhaps one of the most important characteristics of sociology is its scepticism about the possibility of maintaining the fact/value distinction. As far as sociologists are concerned the practice of bioscience and technology is necessarily value laden. This means, that in contrast to bioethicists, sociologists question the increasing social and cognitive authority of medicine, i.e. the acceptance of medical descriptions and prescriptions as objective truths. Whilst the acceptance of the authority of biomedical explanations is not only confined to bioethics, this does not excuse bioethicists' failure to acknowledge the socially or culturally constructed nature of biomedical "facts".

The bioethicists' acceptance of the authority of medicine is demonstrated in current debates about behavioural genetics. We frequently hear pronouncements from the scientific community that, given enough time and money, genetics will enable us to eradicate a range of behavioural "problems" such as gayness, obesity, violent tendencies, alcoholism, and general anti-social behaviour from society. The bioethical debate in this instance centres upon the rights and wrongs of using genetic technologies in this way, - namely the pros and cons of manipulating the human gene pool so as to alter behavioural traits, instead of questioning the socially constructed nature of many of these problems in the first place. In other words, what is frequently overlooked in these debates is that what society considers "acceptable or problematic behaviour", changes both historically and culturally, and is ultimately defined by both political and economic interests.

To give a recent example of the social construction of biomedical "facts" for economic gains, *The Guardian* recently carried a story about the marketing of the anti-depressant Paxil. Paxil is marketed by GlaxoSmithKline, in direct competition with Eli Lilly's 'Prozac' and Pfizer's 'Zoloft'. To attract a larger market share than its competitors, at the same time as Paxil received FDA approval, GSK released magazine articles and national press and TV stories about a new disease - GAD (generalized anxiety disorder) - a "disease" that could be "treated" by Paxil. The story of the discovery of GAD was supported by comments from prominent doctors and psychiatrists, who serve as paid consultants for GSK, and illustrated by dramatic statistics from GSK sponsored studies and patient groups, not unsurprisingly when the news broke, it generated immediate dividends for Paxil. Trade journals and brokers were delighted at the estimate that the strategy would expand the 'anxiety market' to at least $3 billion in the coming years [1]. In effect, GSK manufactured a new disorder in order to market their newly manufactured drug.

It could be argued that, accepting that the scientific endeavour is value laden commits us to a relativistic position which is antithetical to the universalistic principles that underlie many arguments in bioethics. However, we contend that this may strengthen rather than weaken bioethical debate, for as Nelson argues:
'Allowing that scientific enquiry incorporates values is not to say that it is self defeating, but rather to alert us to its complexity' [2]

We similarly believe that bioethics needs to acknowledge these complexities. Bioethicists need to recognise, and have a proper debate about, the role of biotech, pharmaceutical and medical interests in the medicalisation of life. Indeed, it can be argued that unless bioethics, as a discipline, starts to take into account the socially constructed nature of "disease", and the interests that seek to maintain such constructions, it could be accused of rubber stamping biotech's worst excesses.

2) The analytic framework used within bioethics decontextualises ethical issues and does not allow for the flexibility of positions that complex moral problems require.

Whilst we do not deny that moral philosophers may be best qualified to tackle bioethical problems given their training in rigorous reasoning, the primary emphasis on a philosophical methodology within contemporary bioethics, particularly one based in the Anglo-American analytic tradition, can be seen as problematic for the following reasons. First, it overlooks the fact that the questions that are so hotly debated in bioethics are not necessarily hypothetical in origin, as in traditional moral philosophy, but arise within the world, and thus, are generated by particular social and cultural factors and influenced by both political and economic interests. Second, those philosophers who practice bioethics, in contrast to traditional moral philosophy, are increasingly involved in advising policy makers, and as such, their deliberations and conclusions have a direct impact upon the world. Dan Brock argues that there is a 'deep conflict between the goals and constraints of the public policy process and the aims of academic scholarly activity' [3]. He states that:

'The central point of conflict is that the first concern of those responsible for public policy is, and ought to be, the consequences of their actions for public policy and the persons that those policies affect' [4]

The abstract nature of philosophy's methodology is the most frequent criticism leveled at bioethics. To achieve the clarity and consistency required of moral philosophy, broadly accepted analytic frameworks and principals such as those of informed consent and autonomy are used within bioethics to assess the acceptability of particular decisions and actions. These, supported by the meta-ethical foundations of consequentialism, deontology and virtue ethics, allow bioethicists to explore the dilemmas we face from a range of perspectives. These ethical frameworks are necessary to keep hold of a certain objectivity in order that deliberation does not lapse into opinion and preference. However, the flip side of this is that the sparse nature of these analytic frameworks blinds bioethicists to the contextual components that generate moral dilemmas in the first place.

Whilst recontextualising bioethics by bringing the world back in might overcome some of these criticisms it may not necessarily be the answer; for it is not clear to us that the frameworks and modes of thinking involved in analytic philosophy, even supported by empirical data, are comprehensive enough to approach the problems that bioethicists are required to answer. Paradigm shifts not only require the adoption of new methods, but also a new conceptual schema, one that generates new questions. And this leads us to wonder whether bioethics, as currently practiced is really asking the important questions of our time. In other words, in a broadly consequentialist world, do bioethicists really explore the relevant consequences?

The sparse and abstract methodology of moral philosophy tends to define and segregate problems in a way that does not 'fit' onto how most people perceive them. To give a recent example of how public concerns about biotechnology have been misinterpreted, the Nuffield Report on Bioethics on the introduction of genetically modified crops suggested that the public's reluctance to accept this technology was based on their ignorance of risk assessment. However, recent research suggests otherwise, namely, that the public's reactions focused upon questions such as why do we need GMO's and who benefits from any risk that the government wish to make us take, rather than what are the risks and do I wish to take them? [5] In other words, public responses stemmed from a growing mistrust in bioscience rather than a lack of understanding of the risks per se.

3) There is a general lack of consensus concerning the aims and objectives of bioethics. Bioethicists have been criticised for having no coherent end goal and no clear idea of whose ends they are, or should be, serving.

There are many definitions of bioethics, but it is generally agreed that one of its goals is to 'protect vulnerable patients and research subjects from harm and to establish their moral and legal rights'[6]. However, many commentators feel that this is exactly the group that have been let down, in practice. If we look at the United States where the growth in bioethics as an industry, and its acceptance and inclusion in policy matters, has been dramatic, there is a reverse correlation with the availability and quality of health care. Approximately 45 million Americans are now uninsured. There are much greater disparities in life expectancy and treatment availability for those who are particularly vulnerable [7]. These disparities are there for all to see, but as Larry Churchill points out, the deeper problem is the fact that this issue fails to engage us:

'Little notice has been given to the way bioethics is being influenced by the commercialisation in health care. The language is changing to providers and consumers...Market commercialisation and the commodification of health services are in full ascendancy. Do we think that doctors can lose their professionalism but that bioethicists will not, that one can become commodified but not the other?[8]

There is no debate concerning how pharmaceutical companies' marketing budgets are now exceeding their research budgets [9]. No debate on the monopoly of these companies as they continue to merge their interests, or on the question of why the drugs cost so much. Until bioethics is clear about its aims, then, arguably, the question of what types of empirical data to use, and how these may be incorporated within the bioethics project, may be less relevant.

4) The attempt to systematise bioethics has resulted in the reification of autonomy and informed consent as the prime moral goods without adequate reflection on their limitations and proper debate about whom this actually benefits.

The criticisms of principlism (the four principles, justice, autonomy, beneficence and non-maleficence) are well known; for example, that the principles compete rather than complement each other and that autonomy has come to be seen as an overarching principle primarily due to the ease with which it can be "codified" and "regulated". The fact that beneficence and non-maleficence work on the level of the general rather than the particular, plus the existence of competing theories of justice and, and perhaps more importantly, the fact that justice in health care has ceased to be much of an issue, have all contributed to the elevation of autonomy as *the* guiding principle. Even Childress, one of the founders of principlism, has criticised the primacy of autonomy as a moral good [10].

One of the problems with this is that bioethicists do not really address whom this reliance on autonomy advantages. The shifting of responsibility for health from the state to the individual has questionable benefits for "autonomous" patients. Indeed, the beneficiaries of the increasing focus upon autonomy in health care decisions are more likely to be individual health care providers, who are able to use autonomy and informed consent as a method of litigation avoidance, than their patients. Arguably, the consumerisation of biomedicine, which has at its heart the construction of the patient as autonomous consumer of health care choices, has stifled critique within bioethics, and perpetuated inequalities in health care [11]. Whilst institutions and the government avoid overtly interfering in the content of medical decision-making, the creation of funding strategies and capitation schemes limit patients' medical options.

Furthermore, while a preference for autonomy may be a laudable goal, this ignores the fact that in practice the ideal of "free choice" is socially constructed and situated. The existence of social, political and economic constraints, for example: the values inherent within medicine as a discipline; deference to expert knowledge; class differences; ethnicity; education; economic circumstances; religious factors; work needs and the power and influence of the medical profession, serve to undermine the ideal of individual freedom necessary for truly autonomous decisions. Principlism has served as a heuristic device that is accepted, and rarely questioned within bioethics as a discipline. Chadwick and Levitt write that:

'The increasingly influential role of 'the four principles'... in guiding clinical decisions may shed light on the notion of 'ethics as mantra': some suggest that rather than assisting clinicians in clarifying their possible courses of action these principles simply tend to reproduce 'normative frameworks' instead [12].

As noted above, the uni-dimensional view of autonomy which some moral philosophers use, fails to acknowledge the many constraints on autonomy that are recognized by sociologists. This promotes Nelson to suggest that the discipline of bioethics might derive benefit from the adoption of a more social scientific perspective. He states that:

'The social sciences might make a contribution to bioethics by helping the field's practitioners understand better what's behind its deeply installed respect for individual autonomy and whether it has assumed more the character of an ideology than a moral philosophy' [13].

Paul Wolpe notes that the four principles as 'principles embedded in the common morality' were adopted with the goal of pre-empting relativity of judgement. However, he goes on to observe that 'Principles based on a common morality cannot be used to critique that common morality', and argues that the structural factors (for example, professional interests; consumer/capitalist medical structures; the technological and pharmaceutical industries influence and general inequalities of power) which have influenced the adoption and maintenance of these principles should actually be the *subject* of ethical inquiry, not its basis.

5) A limited number of problems are subject to bioethical debate/analysis. There is a lack of reflection about why only certain things present themselves as bioethical problems or dilemmas.

This brings us on to the next point in our list of critiques of contemporary bioethics: namely that bioethical enquiry is limited in scope. Whilst there may indeed be structural reasons for the current focus of bioethics, this does not excuse bioethicists for appearing to ignore some of the more fundamental problems that exist within contemporary health care systems.

As Jeanne Guillemin points out, because of the nature of their work, most referrals for bioethics consultation come from institutions. In other words ethics consultants and committees are assigned their work largely by those who pay for it. Whilst this may be "logical" from a business perspective and perfectly proper, it can have a 'dampening effect' on the efforts of bioethicists to address and analyse other types of ethical problems. Lauri Zoloth writes:

'In a large part, we have turned from the difficult debate about justice because of the way the heroic controversy over the Next New Thing continually shapes and funds our discourse. We have been increasingly asked to focus our attention on the wildly interesting and inventive controversies at the frontiers of science, technology and medicine. We are drawn into this discourse for the same reasons that researchers are: the powerful lure of the new, the availability of funding…and, the promise of revolutionary changes in health care' [14]

The current focus of bioethics is, of course, influenced by the fact that controversy about certain issues such as stem cell research, preimplanataion genetic diagnosis, reproductive and therapeutic cloning is fuelled by media coverage, which in turn influences public debate and hence attracts funding. But with this continuous diversion away from the broader issues of justice in health care, more generally, or debate about the role of bioethics and bioethicists in framing health care policy, we could ask 'do bioethicists fulfil the task of fully representing society's interests'? Do they ask the right questions, or, more pertinently, tough enough questions? What questions is it incumbent upon them to ask? Furthermore, if they don't address the broader questions, and take into account the context of a profoundly unjust healthcare system and the economic logic of the medical/technical industries, then who should or who will? Those who are currently asking such questions – mainly sociologists, find it difficult to get their foot in the bioethics door.

6) Bioethics as a discipline is exclusive. It is dismissive of other disciplinary approaches to, or public participation in debate about, bioethical problems. It employs 'Gatekeeping' practices in determining who should be a bioethicist, i.e. who has the right sort of expertise to comment upon bioethical issues.

Whilst we do not question the need for abstract theoretical discussion about bioethical problems, we are less convinced that this is, or should be, the primary goal of bioethics. However, if bioethics is to change its modus operandi then it may need to include other types of experts and non-experts within the debate. At the present time the adoption of more inclusive practices seems is unlikely. Professional associations of bioethicists often have high membership dues and registration fees – some by invitation only – in some instances academic credentials may be insufficient as, in many fields, particularly clinical bioethics, one's connections are necessary for inclusion. Although this is a criticism of any field of expertise, bioethics, perhaps more than any other discipline, can be criticised for the way in which it polices its boundaries. For examples, Raymond DeVries, the medical sociologist, tells of how he attempted to participate in a Bioethics summer camp. He obtained permission to attend, but the bioethicists participating were displeased and called a special session to decide if he could stay. He was allowed to remain on the basis that he would not name any participant, and permit all of the 50 delegates to review and edit any subsequent article he wrote [15].

For some reason, those moral philosophers who currently practice bioethics, as illustrated above, are notoriously suspicious of sociologists and sociological methods. They are similarly suspicious of lay views on moral issues, because of what are perceived as inconsistencies in reasoning and irrational, ill-defined responses. Larry Churchill writes that:
'We in bioethics do, of course, have things to offer, but our training and pre-occupations do not usually provide an understanding of the varieties of moral wisdom that inform the decisions of ordinary people. It would be foolish to underestimate the importance of this gap in our training and experience, and worse still to assume that no such extra-professional wisdom exists' [16].

However, even if bioethicists think that the public may have misinterpreted the "facts", we would argue, that this is no reason to exclude their views. As the academic Sheila Jasanoff notes, any attempts to resolve issues in a democratic, pluralist society requires bioethicists to acknowledge and respect entrenched normative positions [17].

Discussion

The criticisms outlined above overlook, for the most part, bioethicists' use of, or failure to use empirical data. Instead they charge bioethics and bioethicists with having an insular worldview, more particularly, one which assumes a neutral status quo. So what is the solution? Light and McGee argue that: 'Empirical work will show bioethicists the socially constructed nature of the moral issues they are pursuing'[18].

We would argue that whilst this is important, the real value of adopting a more empirical perspective within bioethics, lies not in the adoption of empirical methods per se, but in the adoption of a different type of world-view. Bioethicists need to be more skeptical – they need to ask 'whose interests are we serving?', they need cultivate an interest in the relationship between structure and agency and make the connection between economic and political conditions and behaviour. They need to reflect upon their role and practices within already existing systems, to be critical of their discipline in order to refine it and make it better. With expert knowledge coming under such scrutiny, particularly with the controversial nature and profound importance of the issues bioethicists debate, they need to be able to answer these criticisms or accept the validity of them and change accordingly. Erica Haimes goes further and suggests that bioethicists themselves need to be the subject of sociological enquiry. She points out that:

'...the empirical work of social scientists [must be seen] within their theoretical roots and hence their analytical framework. It is the theoretical framework that provides the future direction for the relationship between the social sciences and ethics: one that puts ethics centre stage as the subject of social science enquiry. This is a quite different focus and approach to work than that of examining a series of substantive topics (euthanasia, abortion, informed consent) to provide empirical data for ethicists to use to refine their normative analysis. The advantage of putting ethics and ethicists centre stage, as the subjects of sociological research, is that they can then be located in a broader socio-historical context, which enables new questions to be asked. (Such as, why are these issues defined as ethical concerns by these people in these times and these places?) [19].

Ultimately, we would argue that it is time for bioethicists to adopt a more sociological perspective. Sociology could show bioethicists how social structures, historical shapings, cultural settings, power structures and social interaction influence their work. It would allow them to see how bioethics, as a discipline, is constrained by 'disciplinary habits, professional relationships, cultural 'ways of seeing', institutional needs, economic demands, arrangements of power and prestige' and the professionalisation of the discipline within the medical marketplace. A historic reading and teaching of bioethics would also demonstrate how autonomy rather than, for example, justice, came to be the prime moral good and answer those critics who accuse our reification of autonomy as an ideology. The adoption of a more sociological world-view would help bioethicists to understand why only some issues (those involving biotech and pharmaceutical interests) are of interest to bioethics and why clinically-based bioethics is still in the thrall of scientific knowledge as the only valid form of knowledge. Careful reasoning and rigorous analysis require full understanding of the context and circumstances of bioethical debate.The type of expert knowledge we need to be moral experts is truly multi-disciplinary and we need to pay much more than lip service to multi-disciplinarity. As Nelson observes:

'In the end, the contribution of the social sciences to bioethics may be to complete a trend in this interdisciplinary field – that is, to tear down the hold on our imaginations of the notion that the world breaks down along just the lines that intellectual disciplines do'. If bioethics could model a way of understanding and resolving human problems that exhibited both a deep engagement with intellectual disciplines and a freedom from their territorial constraints, that might be among its most profound contribution to its culture' [20].

The marriage of bioethics, empirical data and a social science perspective is a mammoth task. To do it well, we need proper debate about what bioethics should be about and whose interests it should be serving. It also requires a more thorough awareness of the influence of political and economic systems upon bioethical debate and an understanding of how policy is generated and used.

Moral philosophers are indeed, correct to say that the world 'as it is' tells us nothing about the world 'as it should be'. However, bioethicists do not inhabit some hypothetical universe, they exist within the "real" world, they are shaped by, and shape, its institutions. If they are seriously concerned about playing a role in the future regulation of genetics and health care then, their stance needs to be rethought. The problems they tackle arise within the world and the decisions they make impact upon it – for bioethics to be effective as a social good, the complexities of this world must not only be acknowledged, but also embraced.

In conclusion, we suggest there is a need for a new approach within bioethics - a more reflexive, critical and inclusive one. We need a bioethics that not only provides answers to some of the questions of our time, but also tries to evaluate the consequences of its conclusions. Do we need moral experts? Undoubtedly, yes. The moral dilemmas we currently face, such as the genetic modification of human and non-human organisms, generate different and hard questions; we need moral experts to provide some answers.

Finally, given the breakdown of trust in experts across the board, for example, scientists; doctors; lawyers; journalists; accountants, financial advisors and CEO's, it is only a matter of time before the public become equally sceptical of the role of bioethicists. This leads us to one final question namely, who would want to be a new moral expert and why? Perhaps this is the question we should begin with.

References

[1] Koerner, Bernard, (30th July 2002),'First, you market the disease...then you push the pills to treat it', in The Guardian
[2] Nelson, James Lindemann, (2000), 'Moral Teachings from Unexpected Quarters', Hastings Centre Report Jan-Feb, p 14
[3] Brock, Dan W (1999), 'Truth or Consequences: The Role of Philosophers in the Decision Making Process' in Singer, P and Kuhse, H, (Eds), Bioethics: An Anthology, Oxford: Blackwell Publishers, pp. 587-590
[4] Brock, Dan W (1999), ibid
[5] Wynne, Brian, (2001), 'Creating Public Alienation: Expert Cultures of Risk and Ethics on GMO'S', in Science as Culture, Volume 10, Number 4
[6] Wolf, in Holmes, HB, (2001), 'When Health Means Wealth, Can Bioethicists Respond?' , Health Care Analysis 9 (2), p215
[7] Zoloth, Laurie, (2001), 'Heroic Measures: Just Bioethics in an Unjust World', Hastings Centre Report Nov-Dec
[8] Churchill, Larry, (1999), 'Are We Professionals? A Critical Look at the Social Role of Bioethicicts' 'Daedalus' Fall v 128 i4, p.253
[9] Zoloth, Laurie, (2001), op cit, p.36
[10] Wolpe, Raul Root, (1998), 'The Triumph of Autonomy in American Bioethics: A Sociological View' in DeVries, R, and Subedi, J, (Eds.), Bioethics and Society: Constructing the Ethical Enterprise, New Jersey: Prentice Hall, Chapter 3
[11] Wolpe, Raul Root, (1998), ibid, p.53
[12] Chadwick, Ruth and Levitt , Mairi, (1996) Comment, In Cultural and Social Objections to Biotechnology, and in Haimes, Erica (2002), 'What Can the Social Sciences Contribute to the Study of Ethics: Theoretical, Empirical and Substantive Considerations' in Bioethics volume 16 Number 2
[13] Nelson, James Lindemann, (2000), op cit, p.15
[14] Zoloth, Laurie, (2001), op cit
[15] DeVries, R and Subedi, J, in the preface of DeVries and Subedi (Eds), (1998), Bioethics and Society: Constructing the Ethical Enterprise' New Jersey, Prentice-Hall
[16] Churchill, Larry, (1999) op cit, p.253

[17] Jasanoff, Shiela, (1995) Science at the Bar: Law, Science and Technology in America, Cambridge, Mass, Harvard University Press
[18] Light, DW, and McGee, G, (1998), On the Social Embeddedness of Bioethics in DeVries R and Subedi J (Eds.) Bioethics and Society: Constructing the Ethical Enterprise', New Jersey: Prentice Hall pp.1-15
[19] Haimes, Erica, (2002), op cit
[20] Nelson, James Lindemann, (2000), op cit, p.15

Section 2

Empirical Bioethics Considered: Critiquing the Use of Empirical Research

What should we do about it? Implications of the Empirical Evidence in Relation to Comprehension and Acceptability of Randomisation

Angus DAWSON
Centre for Professional Ethics
University of Keele

'why don't they just say "we'll do this" instead of "we'll may be doing this or we could do that".' [1]

Introduction

It is routinely argued that we should gain an informed consent to participation in medical research and that this general rule will apply to the specific example of participation in a randomised controlled trial (RCT). However, there is a large amount of empirical evidence that suggests that gaining an informed consent is a lot more difficult than is often thought, and might actually be impossible with large numbers of patients (even when judged to be competent). [2] Some of this evidence relates to gaining informed consent within the context of RCTs and it is this issue I will focus on here – with the main emphasis on the comprehension and acceptability of the concept of randomisation.

I will argue that the existing empirical evidence in relation to randomisation suggests that large groups within the populations tested (as both participants in trials and in hypothetical situations) seem to have a problem understanding the purpose of randomisation and are apparently keen to reject it as a justifiable methodology. These findings should be of concern to anyone who thinks that RCTs provide strong methodological benefits and also be of concern to anyone committed to the idea that we must gain informed consent before participants are enrolled in medical research. This paper will explore the apparent clash between the empirical evidence on this subject and the requirement to gain an informed consent before participation in a RCT. Various options will be outlined and discussed. It is concluded that we should not ignore this empirical evidence and that this means that we must modify our attitude towards the role of informed consent in a RCT context.

1. Explaining Randomisation

Randomised controlled trials (RCTs) are considered by many to be the 'gold standard' in terms of gathering evidence for medical interventions because of the rigour of the methodology employed. RCTs are used to compare two or more interventions (where one of these might be doing nothing) so that the impact and outcomes can be compared in a systematic way. Randomisation is an important part of the process as it is held that it is an important way to seek to minimise bias. This is because patients are allocated to different cohorts, not by the researchers, which might mean that they could take into account inappropriate factors, but randomly. The power of the methodology is that it is dependent upon this element of randomisation.[3]

Why focus on randomisation in this paper? It might be argued that a discussion centred on this particular concept is unfair because it is one of the most difficult of the concepts employed in a research context to understand. However, I would argue that it is a good example to consider, as it is one of the main reasons why RCTs are so useful and it is a central component of the RCT methodology. If participants do not understand the concept, or its role in the RCT methodology, or accept these benefits, then we can question whether the participant has given a genuinely informed consent to participation in the research. Anyone fully committed to the need to always gain an informed consent [4] in medical research ought to be troubled by this evidence. It is interesting to note that there has been very little discussion about the implications of such evidence for the process of obtaining informed consent within the context of RCTs.

2. The Requirement for Informed Consent

The general consensus in research ethics is that an informed consent should always be gained from competent research participants. Most of the ethical guidelines relevant to research ethics clearly stipulate that this process should include information about the methodology of the research and occasionally randomisation will be explicitly mentioned as the type of thing that participants are supposed to understand. For example the *Helsinki Declaration* states that every potential research participant:

> '*must be adequately informed* of the aims, *methods*, sources of funding, any possible conflicts of interest, institutional affiliations of the researcher, the anticipated benefits and potential risks of the study and the discomfort it may entail......*After ensuring that the subject has understood the information*, the physician should then obtain the subject's *freely-given informed consent*, preferably in writing'. (Paragraph.22)[5]

Likewise the similar ethical guidance from the Council for International Organizations of Medical Sciences (CIOMS) published in November 2002 is even more explicit and detailed as follows:

> '*Before requesting* an individual's consent to participate in research, *the investigator must provide the following informatio*n, in language or another form of communication that the individual can *understand.....for controlled trials, an explanation of features of the research design (e.g., randomization, double-blinding), and that the subject will not be told of the assigned treatment until the study has been completed and the blind has been broken*'. (Guideline 5)[6]

The most obvious justification for seeking an informed consent within the context of medical research is by appeal to the principle of respect for autonomy.[7] The idea is that the participant must understand all of the information that might be relevant to a decision to participate or not. Only if all such information is given and understood can the person be held to have genuinely given an informed consent. Part of the reason for such an emphasis upon the idea of consent in a research context arises from the wish to avoid a repeat of the historical examples of research abuse. [8]

3. Empirical Evidence

So far we have seen that the methodological benefits of RCTs are clear and that there is a moral requirement to provide potential participants with the relevant background for an informed consent. This is taken to mean that we should provide all relevant information and ensure that the participants understand it. However, we cannot simply end discussion at this point because there is now a significant amount of evidence that patients find it hard to comprehend and/or accept a number of the elements of the RCT. I will concentrate here on a number of studies that focus on lay understandings of randomisation and its acceptability as part of a research methodology.

A number of related studies reported in two papers by Appelbaum, Roth and Lidz (1982) and Applebaum, Roth, Lidz, Benson, and Winslade (1987)[9] outline what they term the 'therapeutic misconception' that they have observed amongst many of the participants in RCTs involving psychiatric treatments. This misconception arises because participants in research often believe that they are being treated on the basis of what is in their individual interests rather than according to a randomised research protocol. This belief then provides an explanation as to why participants fail to see the research context as being different from the clinical one. On this view, doctors and nurses are seen by the patients to be conducting the research so it makes sense to patients to think that the actual intervention they will receive has been chosen because it is in their best interests. They believe this despite having been given information about the nature of the research and being held to have given an informed consent to participate in the research. The results of one study suggest that 69% of participants interviewed had 'no comprehension of the actual basis for their random assignment' and only 28% were held to have 'a complete understanding of the randomisation process'. [10] Appelbaum et al. hold it to be the job of researchers to seek to eradicate these incorrect assumptions where they exist. However, Robinson et al. argue that:

> 'Disabusing patients of incorrect assumptions may not however be sufficient to ensure they understand and accept the true situation' [11]

This is certainly true. However, presumably Appelbaum et al. are suggesting that, in their view, patients' assumptions can be 'corrected' through the provision of more information, or if information is provided in a different form, or by those specially trained to do so. However, we might doubt this conclusion if we consider the issues against the backdrop of the following studies as well.

Snowdon, Garcia & Elbourne's study [12] explored parental reactions to random allocation of treatment within a trial for neonates with breathing difficulties. The two arms of the trial were standard treatment using ventilatory support versus oxygenation of the blood via an external circuit (ECMO). The study involved in-depth interviews with 37 parents some months after first participation in the trial. All parents were held to have given informed consent before enrolment into the trial. However, the results suggest that the parents frequently misunderstood the concept of randomisation and the role of randomisation in the methodology of the trial. For example, many of the parents apparently saw randomisation as being related to the issue of *access* to the trial (and therefore a chance for their child to get the new treatment) rather than about the *methodology* of the research. The paper provides an analysis of the different 'models' provided by the parents as explanations for the events as they occurred. These models make clear that many of the parents were constructing their own alternative accounts of events in an attempt to justify the role that randomisation played in relation to their child's situation. Common examples of this included the suggestion that staff provided the treatment that they believed was appropriate therapeutically for the child, whereas others suggested that the actual outcome for the individual child (standard treatment or ECMO) was linked to the need to ration a limited resource. Both of these explanations suggest that parents did not understand the role of randomisation and its methodological justification in the RCT.

Similar results were obtained in the study by Featherstone & Donovan. [13] They argue 'that participants engage in an ongoing struggle to understand the methods of the trial and the process by which they are allocated treatment'. [14] This study focused on in-depth interviews with 33 middle-aged and older competent patients with common urinary problems offered the opportunity to participate in a series of linked trials comparing three different treatments (a new treatment (laser surgery); standard surgery (transurethal resection of the prostate - TURP); and conservative management (monitoring but no active intervention)). The authors suggest that the effort of the participants to comprehend their involvement in this RCT led them to adopt 'several approaches to making sense of the trial'. [15] and that in many cases these differed radically from those of the researchers. One of the main problems the participants [16] had was with the issue of randomisation. Featherstone and Donovan state that:

'Just over half of the participants (12) indicated that they had expected to receive treatment based on their diagnosis and an assessment of their specific needs by a clinician […] in the way that they perceived normal clinical practice to occur' [17]

All participants were held to have given an informed consent to participation in the trial. However, here, as in the studies by Appelbaum et al. and by Snowdon et al., we can see the influence of beliefs about the nature of 'normal' clinical practice at work in the background. The participants' misunderstanding was, apparently, re-enforced by the use of tests and questionnaires to determine if they met the entry criteria for the trial. Many of the patients saw these as being about determining what was best for the patient rather than about 'eligibility' for entry to the randomised trial. The degree of misunderstanding of the nature of the RCT is perhaps best represented by the fact that many of the participants expressed a clear view about which of the alternatives they would have preferred. Many of those not getting what they wished for felt cheated. Such an outcome confirms the problems in comprehending the nature and application of randomisation.

In a forthcoming study Robinson, Kerr, Stevens, Lilford, Braunholtz & Edwards report a series of studies into randomisation (using medical and non-medical hypothetical examples) with non-trial participants. The subjects for the studies were all adults attending adult education classes. The broad results were that most participants did not consider allocating treatment at random to be acceptable as a methodology in a medical or non-medical context. It is clear that the scientific benefits of randomisation were not considered by the participants to be sufficient to justify the use of a randomisation element within the context of research. For example the discussion of the second of four studies suggests that:

'The results give no indication that participants appreciated the scientific benefits of random allocation of treatments in a clinical trial over patient choice'. [18]

Robinson et al. conclude that:

> 'participants often found it unacceptable for a clinician to suggest deciding between treatments at random'

and also that:

> 'participants assumed that just as much knowledge would be gained about which treatment is better if patients and doctors chose their treatment, rather than if treatment were allocated at random.' [19]

Having outlined some of the relevant studies we can attempt to consider in broad terms what the evidence in relation to randomisation suggests. We can separate out different issues. Do participants in RCTs understand the concept of randomisation? Do they see the reasons for its deployment? Do they accept it as a legitimate methodology with clear scientific benefits? The answers to these questions, based on the research as reported above, are that a substantial number of people (perhaps as high as half of the groups studied) of those judged to be competent and judged to be able to give an informed consent do not understand how a randomisation methodology works or can see the benefits of using it.[20] In addition, the work of Robinson et al. suggests that a majority of potential participants are likely to question whether the use of a randomisation methodology is actually ethical. Perhaps, we should not be surprised about these findings in relation to randomisation, as they are, in fact, broadly consistent with 30 years of empirical research in relation to the comprehension of informed consent in general. [21]

4. Methodological Reflections on this Evidence

The studies reported above jointly suggest that research participants have a great deal of difficulty in comprehending and accepting the idea of randomisation. Of course in drawing conclusions from such studies we need to be cautious. Despite such a caveat, the only fair conclusion is that we have good grounds for arguing that a large number of competent people who are apparently able to give an informed consent to participate in an RCT do not really fully comprehend what randomisation involves or can see its methodological benefits. This is a startling and important conclusion.

On the other hand, we always need to take care not to generalise from one research context to another. In the light of this principle there are a number of issues that should be noted. For example, in considering the evidence we need to be clear about whether the studies relate to research within actual trials or hypothetical trials. Doing research within actual trials is almost certainly preferable as we are closer to the actual situations we are interested in. However, the participants, for the same reason, are more likely to be anxious and stressed, particularly if the condition studied is in an emergency neonatal context such as in the study by Snowdon et al. The advantage of hypothetical trials or research outside of a real RCT is that the results might relate more 'purely' to the research question studied. However, the disadvantage is that it is not clear that we are studying the same thing as in the real RCT situation (given the potential complicating factors as described above). [22] Perhaps the best situation is a mix of the two methodologies and then we can compare results and consider whether there is anything significantly different between the two and then, in turn, this can be explored. There are a number of other methodological issues that are worth careful consideration. For example, many of the studies of actual RCTs involve the investigation of information recalled by the participants at some later date and not at the time of the consent to the procedure. These interviews occur in a timeframe from the next day to a few months later. [23] Another issue is that some studies involve research questions focusing upon participants' attitudes towards randomisation rather than participants' understanding of it. A further important issue is that the studies related to real RCTs use qualitative methods such as interviews and consequently the size of the population studied tends to be quite small. This does not, of course, invalidate the studies, but it does suggest that we need to be cautious in generalising from the results. However, it is significant that broadly similar results were produced by all three studies within existing trials [24] despite the diversity of treatments looked at (psychiatric research; neonatal respiratory problems; common urinary symptoms in middle aged and aged men).[25]

5. Process Failures and Framing Failures

Where evidence exists of misunderstandings or lack of comprehension the obvious suggestion is that this failure will be because the information is provided in a way that is too complicated for the potential participant or that insufficient time was given to the informing process by the researchers. If these really are the failings then they might be ameliorated by modifying the informing process itself. Let's call such failures, 'process failures', as they emerge from a failure in the informed consent process to provide the best possible environment for a truly informed consent. Such process failures are bound to occur at times given all the potential problems involved in communication between individuals but the empirical evidence taken together suggests that a different type of failure is also present. The latter failures are due to the way that the information accepted is actually influenced by an individual's background or pre-existing beliefs and attitudes. When an individual has a failure to understand on this basis we can call these 'framing failures' since the errors occur not because the information is lacking or has been misunderstood but because it is taken in against the framework of an existing web of beliefs, experiences and attitudes. These prior states 'frame' the way that the new information is comprehended. [26]

Despite the startling results of the empirical studies above suggesting this 'framing', the discussions and conclusions of the papers largely focus on 'process' failures. Each study contains an almost ritualistic appeal to the need for more information and more time for the informing process, as well as the use of different methods to provide the information. For example, Snowdon et al. suggest that there are other ways to give information that may be useful [27] and Featherstone & Donovan suggest the need for patients 'to discuss the reasons for particular methods of trial design (such as randomisation) with researchers and reflect on these in order to give true informed consent'. [28] Robinson et al. argue that in a real trial situation participants' assumptions will remain different from those of the researchers 'unless additional explanatory information is provided.' [29]

Occasionally in these studies it is hinted that the provision of information or adaptation of the information-giving process might not be enough, such as when Featherstone and Donovan admit that:

'Although this study confirms the importance of providing clear and accurate patient information, it also shows that this in itself is unlikely to ensure consistent interpretation of concepts such as randomisation by participants' [30]

In their later paper they conclude that:

'Such interventions require further research, but the findings from this study suggest that most participants (and non-participants), whatever their level of knowledge, will struggle to make sense of the need for randomised trials.

Perhaps the greatest need is for more open debate about trials amongst trialists, recruiting clinicians and the public' [31]

It is not clear what the last sentence refers to, but presumably it places emphasis back upon looking for ways to work through improving the information giving process through public debate despite the findings of the paper itself, suggesting that a deeper problem exists in relation to randomisation. Notwithstanding that the conclusions of the empirical studies all tend to focus on the idea that the provision of more information or the supplementation of information will provide the solution to process failures, there is no reason to think this is the case. [32]

Another way to interpret these studies is that the 'framing failures' that they uncover are a serious problem that cannot just be "informed away" in the way that is often suggested. This conclusion is one informed by the available empirical psychological research about the nature of human beings as consumers of information. [33] On this view, the deeper problem may lie not in the fact that participants cannot understand the information in the sense that there is no information given to them that is in itself incomprehensible; it is just that individuals are not mere recipients and processors of pure information. The fact is that we do not just absorb information in a neutral and purely rational way even if it is presented as well as it possibly could be. We interpret, judge, order and shape that information according to our prior beliefs, assumptions, social situation et cetera. Often we will struggle to make sense of something because of the way we already view a particular situation. Where situations are complex we tend to provide our own interpretation or explanation about what is happening so that it fits better with the 'facts' as we see them: such traits can be seen in these studies. This insight provides an explanation about such things as Appelbaum et al's 'therapeutic misconception', the findings of Snowdon et al., and those of Featherstone and Donovan, in relation to the reluctance of participants to see the various treatment options as being truly allocated at random. The idea of 'framing failures' takes into account these types of psychological processes. This view is not supposed to suggest that such patients are either stupid or incompetent. It is not that participants failing to comprehend randomisation have some sort of 'deficit' of understanding that can be made up by the provision of even more information. The view is, rather, based upon observed experience about the empirical facts relating to the way that humans tend to process information. Of course, this does not mean that we should not try to inform participants about randomisation, but it does raise the question of the degree it is possible to achieve such an aim. This points the discussion clearly towards ethical issues, but before I consider these in detail, there is another element in one of the empirical studies to bring out as it intimates another possible explanation for why there is reluctance of patients to accept randomisation as a research methodology.

The study by Robinson et al. suggests that large numbers of participants do not only not see the reason for randomisation, they, apparently, also find the use of such a methodology in a piece of research ethically problematic. Of course the participants in this study were not medical patients at the time, and it is just possible their views might have been different in such a context. However, if these findings do reflect something fundamental in the way that many people see randomisation then it is plausible to think that such ethical beliefs might be playing a similar role in the response to randomisation as the other 'framing failures' outlined in the empirical studies discussed above. If this is so, then we end up with a curious set of facts as follows: a group of competent patients give their consent after being informed of different aspects of the research, including the facts of randomisation; and yet, despite the fact that the participants are unhappy with the justification for the methodology of the trial, they still consent to participate. From such a description it sounds as though the patients are just confused. Perhaps this is the case.

However, I don't think we should necessarily draw this conclusion. It seems likely that this is just what our psychology is like, given that our beliefs in fact consist of messy webs rather than systematic and coherent sets. The fact that many people find randomisation ethically problematic suggests either that this leads their misunderstandings of the trial or because they are confused about the methodology they are not willing to give the methodology their ethical endorsement.

6. What Should We Do?

The reason to ask this question is that it looks as though we are faced by an apparent dilemma here. Either we continue to promote the popular ethical principle of respecting autonomy through gaining informed consent to research or we take account of the empirical evidence and revise our practices in this area. [34] I suggest that there are at least three options here which I will label: denial, uphold autonomy, and revision.

(a) Denial

This option involves discounting or ignoring the empirical evidence and carrying on with the current approach. This is an essentially pragmatic response in that it might be supported by the thought that the relevant issue is that competent people do (apparently) give informed consent to participate in research. Therefore, as long as this continues to happen and there are no complaints we don't need to look too closely at the details. However, this approach seems problematic. If evidence of problems in comprehension exist, should we ignore them? Should we carry on pretending to obtain informed consent knowing that, in many cases, this is just a charade? This approach essentially discounts the empirical evidence even if it is felt to be relevant to the attainment of the ethical end. It can be objected that such an approach is complacent and shows a lack of respect for the research participants. On such a view, once a problem is known about it should not be ignored.

However, a possible justification for this method might be that on balance *trying* to obtain informed consent might be enough to justify the research; but this seems incompatible with the spirit of gaining informed consent. The only honest description of such an approach is to say that research participants are informed (in the sense that they are given the information) but there is no guarantee that they understand the information and base their decision upon it. However, support for this approach might come from the fact that patients generally report satisfaction with the consent process. [35] Even if this is the case it might still be objected that respect for autonomy requires something greater than merely going through the motions of gaining informed consent knowing that it is likely to be meaningless in a significant number of cases.

(b) Uphold Autonomy

This second option would take the empirical evidence very seriously and suggest that we need to modify the research process as a result of the discovery of 'framing failures' in this area. The idea would be that informed consent is so important, due to our need to respect patient autonomy, that we either need to, at the very least, limit those eligible to participate in RCTs to those we know do understand the methodology or we consider bringing the use of the RCT methodology to an end altogether. The first view, the more modest version of this option, limits those eligible to participate in RCTs to only those individuals that we are happy can fully understand the details of the methodology employed. [36] This approach will not end RCTs but it will make it much harder to recruit to trials. This, in turn, will mean that fewer trials will be undertaken and those that are begun will take longer before significant results are obtained. The second (stronger) view, that we should consider bringing the RCT methodology to an end, is clearly a possible, but rather desperate, option. The advantage is that it preserves the value that autonomy seems to have come to have in discussions about research. However one objection might be that this is obtained at too high a price. The price is that much useful research will either not be done or will be done using poorer methodologies. My own view would be that this price is indeed too high. Both of these consequences will have a significant negative impact upon patient welfare. Any advocate of this second option, whether of the modest or strong form, must realise this is the result of their insistence upon upholding autonomy and be happy to accept the subsequent reduction in welfare. Of course, many will be happy to do this as they wish to prioritise autonomy in any clash between different ethical values. [37]

(c) Revision

This third option follows from accepting the empirical evidence on consent and the need to do something about it (so rejecting option (a)). The suggestion is that, given the benefits of using a randomised methodology, we also reconsider the importance we give to informed consent in an RCT context and focus on other possible justifications for research rather than solely focusing on autonomy (so rejecting option (b)). We are left with option (c) as being the most plausible.

The basis of this view is that we should take seriously the empirical evidence about 'framing failures' but we also need to accept the methodological benefits of randomisation. The evidence is consistent and forceful in suggesting that a large number of people involved in RCTs do not understand the concept or comprehend the methodological benefits of randomisation. The presence of 'framing failures' rather than 'process failures' means that the misunderstandings and lack of comprehension cannot just be informed away by providing more and more information. This empirical evidence, taken with the appeal of the general moral principle that 'ought implies can' (that is, we can only be obligated to perform an action if it is possible to achieve it) suggests that we do need to reconsider the role of informed consent in the RCT context. The conclusion to be drawn is that if it is indeed impossible to achieve true informed consent in relation to randomisation then we should not be morally required to attempt to do so. In the potential conflict between what is morally required of us as researchers (as insisted upon by Helsinki and CIOMS) and the empirical evidence about the practicalities of achieving those ends, the ideal of informed consent must be compromised as a result of empirical reality. If this is true then here is an example of a situation where the normative is, or ought to be, constrained by the empirical evidence.

The practical conclusion to be drawn is that we should revise our attitude towards informed consent in research and move to an approach that puts greater weight on other considerations such as beneficence or welfare issues. It is often forgotten that participation in an RCT might have various potential benefits for the patient themselves as well as the rest of society. A good example of this is provided by the ECMO study that was the basis of the paper by Snowdon et al. The newborn babies with breathing difficulties were randomised to one of two possible treatments. One of these was the standard treatment the child was receiving already; the other was the new treatment (ECMO). Given the fact that the entry criteria for the trial required that the children were seriously ill on the existing treatment, arguably, the children could be no worse off trying the 'experimental' treatment. These welfare considerations should be the basis of the research ethics committees' deliberations rather than excessive concerns about whether patients (or their parents, in this case) comprehend all aspects of the research methodology. Parents naturally want what is best for their child and in this particular case this in all likelihood involves enrolment in the trial. [38]

Conclusions

More research into the problems of gaining informed consent in different research scenarios and with different patient groups should certainly be performed, and there should certainly be more work done on following up the work of Robinson et al. and their interesting findings about the acceptability of the randomisation methodology. However, the existing empirical evidence reviewed in this paper provides a clear message for anyone willing to see it. It suggests that obtaining an informed consent in relation to randomisation in this context is likely to be impossible for a sizeable number of people (perhaps as much as half the population). On the other hand, we are surely justified in conducting such research in at least some contexts given the methodological advantages. I believe that the focus should move from an obsessive concern about the role of informed consent to an assessment of the balance of harm and benefits of the research in that particular context. Those able to perform a protective role for patients, such as research ethics committees, should conduct this assessment. Consent (in general terms) can still be given by individual participants before enrolment in any trial, but we should not worry unduly about the fact that the relevant patients do not understand and/or endorse every aspect of the research methodology involved. Competent research participants should consent but not necessarily give an informed consent before enrolment in a trial. [39]

References

[1] A patient 'Kim' talking about her experience of randomisation as a parent of a child within the 'ECMO trial', quoted in Snowdon et al. p.1344
[2] See Dawson (forthcoming) for a critical review of this literature
[3] Kunz et al. (2003) for a systematic consideration of the benefits of randomisation. Kunz, R, Vist, G, and Oxman, A, (2003), 'Randomisation to Protect Against Selection Bias in Healthcare Trials, (Cochrane Methodology Review)'. In: *The Cochrane Library,* Issue 3, 2003. Oxford: Update Software
[4] Except perhaps where the participant is incompetent or there are other clear exceptional reasons (such as the need to act in an emergency situation)
[5] My italics. World Medical Association. *The Declaration of Helsinki.* Amended 2000, with clarification 2002. http://www.wma.net/e/policy/b3.htm
[6] My italics. CIOMS. *International Ethical Guidelines for Biomedical Research Involving Human Subjects.* November 2002. http://www.cioms.ch/frame_guidelines_nov_2002.htm
[7] Other justifications might also be available such as an appeal to welfare considerations. See Brock, D, (1987) 'Informed Consent', in VanDeVeer, D, & Regan, T, (eds.) *Health Care Ethics: An Introduction.* Philadelphia: Temple University Press
[8] See Brody, B, (1998), *The Ethics of Biomedical Research: An International Perspective.* Oxford: Oxford University Press (especially chapter 2)
[9] Appelbaum, P, Roth, L, & Lidz, C, (1982) 'The Therapeutic Misconception: Informed Consent in Psychiatric Research', *International Journal of Law & Psychiatry,* Vol.5: 319-329; Applebaum, P, Roth, L, Lidz, C, Benson, P, & Winslade, W, (1987) 'False Hopes and Best Data: Consent to Research and the Therapeutic Misconception', *Hastings Center Report,* Vol.17: 20-24, See also Berg, J., Appelbaum, P, Lidz, C, & Parker, L, (2001), *Informed Consent: Legal Theory & Clincial Practice (2nd ed.).* Oxford: Oxford University Press
[10] Applebaum, P, Roth, L, Lidz, C, Benson, P, & Winslade, W, (1987) 'False Hopes and Best Data: Consent to Research and the Therapeutic Misconception', *Hastings Center Report,* Vol.17: 20-24, p 21
[11] Robinson, E, Kerr, C, Stevens, A, Lilford, R, Braunholtz, D, Edwards, S, (forthcoming) 'Lay Conceptions of the Ethical and Scientific Justifications for Random Allocation in Clinical Trials', *Social Science and Medicine,* p7

[12] Snowdon C, Garcia J, Elbourne D (1997), 'Making Sense of Randomisation; Responses of Parents of Critically Ill Babies to Random Allocation of Treatment in a Clinical Trial. *Social Science and Medicine*; 45(9): 1337-1355

[13] Featherstone, K, Donovan, JL, (1998) 'Random Allocation or Allocation at Random? Patients' Perspectives of Participation in a Randomised Controlled Trial', *British Medical Journal*, Vol.317: 1177-80; Featherstone, K, Donovan, JL, (2002) '"Why Don't They Just Tell Me Straight, Why Allocate It?" The Struggle to Make Sense of Participating in a Randomised Controlled Trial', *Social Science and Medicine*, Vol. 55: 709-719

[14] Featherstone, K, Donovan, JL, (2002), ibid, p.718

[15] Featherstone, K, Donovan, JL, (2002), ibid, p.717

[16] As did the non-participants: the study, unusually, includes interviews with both groups

[17] Featherstone, K, Donovan, JL, (2002), ibid, p.713

[18] Robinson, E, Kerr, C, Stevens, A, Lilford, R, Braunholtz, D, Edwards, S, (forthcoming) 'Lay Conceptions of the Ethical and Scientific Justifications for Random Allocation in Clinical Trials', *Social Science and Medicine*, p19

[19] Robinson, E, Kerr, C, Stevens, A, Lilford, R, Braunholtz, D, Edwards, S, (forthcoming), ibid, p29

[20] The studies conducted by Robinson et al. also suggest that many people might even object to it on ethical grounds (some might be tempted to argue, controversially, that the fact that they do not approve also demonstrates that they don't understand). These findings fit in with other empirical studies that have been conducted looking at researchers' beliefs about participants' comprehension of randomisation. In general it would appear that researchers are sceptical about the degree of patient understanding about randomisation (See, for example, Taylor, K, Kelner, M, (1987) 'Interpreting Physician Participation in Randomised Clinical Trials: The Physician Orientation Profile', *Journal of Health & Social Behavior*, Vol.28, No.4: 389-400; Verheggen, F, Jonkers, R, Kok, G, (1996) 'Patients' Perceptions on Informed Consent and the Quality of Information Disclosure in Clinical Trials', *Patient Education and Counseling*, Vol.29: 137-153; Ferguson, P, (2003) 'Information Giving in Clinical Trials: The Views of Medical Researchers', *Bioethics*, Vol.17, 1: 101-111. The study by Kerr, C, Robinson, E, Stevens, A, Braunholtz, D, Edwards, S, Lilford, R., (forthcoming) 'Randomisation in Trials: Do Potential Trial Participants Understand It and Find It Acceptable?', *Journal of Medical Ethics*, would, however, seem to provide some evidence to contradict this view. The conclusion of the latter study is that the general public 'understand' randomisation, or rather, they agree with the researchers on paradigm cases. The study also demonstrates that the subjects do not accept randomisation as a methodology. However, we need to be cautious as this study involved non-trial participants and hypothetical cases.

[21] See Sugarman J, McCrory D, Powell D, et al. (1999) 'Empirical Research on Informed Consent: An Annotated Bibliography'. *Hastings Center Report*. Vol. 29: Supplement, S1-S42. for a review and Dawson, A. (forthcoming) 'Informed Consent: Bioethical Ideal and Empirical Reality', in Hayry, M., Takala, T. & Herissone-Kelly, P. (eds.) *Bioethics and Social Reality*. Amsterdam & New York: Rodopi, for a discussion of this literature

[22] This is because such 'complicating factors' might be vital components of the situations studied

[23] Of course, we cannot simply base understanding of 'x' upon recall of 'x' when told of 'x' at one time and asked about it later. As time passes between the point of information and the point of recall we are likely to become more cautious about this. However, if recall is poor it is surely a reasonable assumption that comprehension at the time of information provision was not deep.

[24] Applebaum, P, Roth, L, Lidz, C, Benson, P, & Winslade, W, (1987), ibid; Snowdon C, Garcia J, Elbourne D (1997), ibid; Featherstone, K, Donovan, JL, (2002), ibid

[25] See also Gammelgaard, A. (2003) *Ethical Aspects of Clinical Trials involving Acute Patients – described in relation to the DANAMI-2 trial*. PhD Dissertation. University of Copenhagen, Denmark, for details of research involving patients suffering from myocardial infarction

[26] The name, of course, derives from the substantial psychological literature on this phenomenon. See Tversky, A, Kahneman, D, (1981) 'The Framing of Decisions and the Psychology of Choice', *Science*, 211: 453-458; Tversky, A, Kahneman, D, (1984) 'Choices, Values and Frames', *American Psychologist*, 39: 341-50

[27] Snowdon C, Garcia J, Elbourne D (1997), ibid, p 1350

[28] Featherstone, K, Donovan, JL, (1998), ibid, p.1180

[29] Robinson, E, Kerr, C, Stevens, A, Lilford, R, Braunholtz, D, Edwards, S, (forthcoming), ibid, p.30

[30] Featherstone, K, Donovan, JL, (1998), ibid, p.1179

[31] Featherstone, K, Donovan, JL, (2002), ibid, p.718

[32] Interestingly there is some evidence that where more information is given the number of participants going on to give consent falls. See Wragg, J, Robinson, E, & Lilford, R, (2000) 'Information Presentation and Decisions to Enter Clinical Trails: A Hypothetical Trial of Hormone Replacement Therapy', *Social Science and Medicine*, Vol.51: 453-462. However, we cannot infer that there is any increased understanding as a result of the extra information.

[33] See Kent, G, (1996) 'Shared Understandings for Informed Consent: The Relevance of Psychological Research on the Provision of Information', *Social Science and Medicine*, 43 (10) 1517-1523, for a useful summary of much relevant psychological evidence.

[34] Of course this argument depends upon accepting that the empirical evidence is convincing but even if it is not acceptable, we could always have a hypothetical version of the argument

[35] For example, see Harth, S, Thong, Y, (1995) 'Parental Perceptions and Attitudes about Informed Consent in Clinical Research Involving Children', *Social Science and Medicine,* Vol. 41, No. 12: 1647-1651; Verheggen, F, Jonkers, R, Kok, G, (1996) 'Patients' Perceptions on Informed Consent and the Quality of Information Disclosure in Clinical Trials', *Patient Education and Counseling*, Vol.29: 137-153;

[36] We might well exclude such a group because we judge them to be incompetent in respect of this particular decision. This would depend upon the adoption of a 'decision-specific' view of competence as suggested by Brock and Buchanan (Brock, D, Buchanan, A, (1989) *Deciding For Others: The Ethics of Surrogate Decision Making.* Cambridge: CUP). I'm grateful to Stephen Wilkinson for discussion of this point.

[37] Such a role for autonomy is, in my view, completely implausible but this is not the place to argue for this.

[38] This is not to say that participation in a trial will always be so clearly of benefit to the participant as it is, arguably, in this case. This is just an example of how the welfare considerations might be drawn in one particular example. However, if the research community is genuinely in equipoise in relation to the effectiveness of the proposed interventions (even if the RCT involves active treatment versus placebo), then at the very least patients have equal grounds for thinking they may or may not benefit. If the RCT involves existing treatment versus new treatment they perhaps have stronger grounds for hoping for greater benefit.

[39] I am grateful to audiences at the second annual "Northweb" seminar entitled 'Bioethics and Social Reality' held in Manchester, UK (February 2003) and the XVIIth ESPMH Conference in Vilnius, Lithuania (August 2003) where I gave earlier versions of this paper. Thanks also to my colleagues at Keele University for a number of stimulating discussions on this topic and to Mike Dawson for commenting on the (almost) final version.

Use and Abuse of Empirical Knowledge in Contemporary Bioethics: A Critical Analysis of Empirical Arguments Employed in the Controversy Surrounding Stem Cell Research *

Jan Helge SOLBAKK
*Center for Medical Ethics, Faculty of Medicine,
University of Oslo*

Introduction

In two forthcoming articles about the controversy surrounding stem cell research, [1,2] Søren Holm claims that no argument has so far been advanced in the debate to justify the *necessity* of destructive research on human embryos in order for the therapeutic potential of stem cell research to be achieved, and that it is up to the scientists themselves to produce 'convincing arguments' for their case [3]. This seemingly defeatist statement on behalf of bioethics originates from the viewpoint that neither a reiteration of old arguments about the moral status of the human embryo nor the generation of new arguments of the same kind are likely to have any positive bearing upon the controversy; on the other hand, the impact of science on the current debate is unquestionable, due to three 'partially independent' developments:

1) the invention of methods for derivation and in vitro cultivation of human embryonic stem cells (ES-cells),

2) the invention of techniques for cell nuclear replacement (CNR), and

3) the publication of several studies suggesting that adult stem cells are in possession of a much higher *plasticity* than previously believed [4].

This paper has three aims. The first is to identify different forms of empirical arguments employed and to critically assess their function in the debate. The second aim is to show that not only is there insufficient scientific evidence available to therapeutically justify human embryonic stem cell research; the same holds true for the opposite case, i.e. of using therapeutic arguments to question the necessity of human embryonic stem cell research. Finally, I want to draw attention to a set of empirical arguments that seem to meet Holm's requirements of justification for allowing destructive stem cell research on human embryos. In order to facilitate the differentiation of empirical arguments employed in the debate, I will rely on the categorisation of research in a report on stem cell research issued by The Select Committee, House of Lords, United Kingdom on February 13th 2002 into [5]:

Types of Empirical Narratives Employed in the Debate

The overstated way of story-telling

Holm's main critique against the protagonists of embryonic stem cell research is that their position is based on unwarranted empirical premises [6]:

'The benefits that are put into the balance to justify the sacrifice are mainly the therapeutic potential promised by stem cell therapy. The public presentation of the benefits of stem cell research has often been characterised by the promise of huge and immediate benefits. Like with many other scientific breakthroughs the public has been promised real benefits within 5-10 years, i.e. in this case significant stem cell therapies in routine clinical use. Several years have now elapsed of the 5-10 years and the promised therapies are still not anywhere close to routine clinical use'.

A typical example of this kind of argumentation is found in the document of Opinion, 'Ethical aspects of human stem cell research and use' of the European Commission's Group on Ethics in Science and New Technologies: [7]

'The Group notes that in some countries embryo research is forbidden. But when this research is allowed, with the purpose of improving treatment of infertility, *it is hard to see any specific argument which would prohibit extending the scope of such research* in order to develop new treatments to cure severe diseases or injuries. As in the case of research on infertility, stem cell research aims to alleviate severe human suffering.'

In their account the Group seem to take for granted not only that the motivation behind embryonic stem cell research is of the same alleviating kind as embryonic research to improve treatment of infertility, but also that the current situation of estimating the benefits of embryonic stem cell research is comparable to the situation at the time when embryonic research aimed at improving treatment of infertility was introduced at the end of the 1970's. Such a comparison is, however, hardly warranted: At the time when embryonic research to improve treatment of fertility was introduced, IVF was already an existing form of treatment, albeit with a lower success-rate than quantifiable estimates suggested that it was possible to achieve by allowing methodological research on human embryos. Embryonic stem cell research, on the other hand, is still in an 'embryonic state' - to use Holm's expression - still far away from any possibility of quantifying its therapeutic potentials. Or, to put it more bluntly: The empirical basis for using the 'beneficial' sacrifice of human embryos involved in research aimed at infertility treatment as a justification for extending the scope of destructive research to include embryonic stem cell research simply does not exist. Consequently, the Group's inference from analogy also has to be rejected.

Fact-Finding Narratives

Although there are numerous other examples in the literature of similar attempts at therapeutically overselling embryonic stem cell research [8], this does not represent the whole story. Alternative 'scientific' narratives are represented in the debate as well and deserve more moral attention than hitherto given. One of the most recent of such accounts is to be found in the report on Stem Cell Research issued by the House of Lords Select Committee, mentioned above. What makes this report a different reading from many other contributions is that priority is given to identifying various forms of *basic* stem cell research that need to be undertaken *before* anything can be said definitely about the possibilities of performing pre-clinical studies and/or large-scale clinical trials. No attempt

is made here at 'promising too much too early' – to again use Holm's formulation. On the contrary, the authors of the *Report* are cautious to get the vital facts as straight as possible and to bring speculations down to a minimum: [9]
'The great majority of potential stem cell-based therapies are still at the first stage of this process, basic scientific research' (2.11).

'From the evidence we have received we are clear that over the next few years most studies on stem cells, whether adult, foetal or embryonic in origin, will be basic research. This research will not in itself be therapeutic, but will be undertaken with the aim of gaining the understanding necessary if stem cells are to be used widely for therapeutic benefit' (2.12).

In the Select Committee's *Report*, nine basic research challenges are identified as necessary to explore prior to attempts at performing pre-clinical or clinical studies on living individuals:
1. Identification and characterisation of stem cell specificity.
2. Isolation and purification of specific stem cells in 'sufficient number to be useful'.
3. Creation of *clean* growing conditions in vitro.
4. Maintenance of stability of acquired properties after undergone manipulation.
5. Directed stem cell differentiation.
6. Controlled stem cell integration.
7. Proper understanding of the processes of de- and re-differentiation.
8. Control of stem cell migration.
9. Avoidance of immunological rejections.

Competing Stories of Redundant Research

In its response to the question whether research on human embryonic stem cells is *necessary* in order to cope with these challenges, the Committee rely on a set of empirical premises worth closer examination. The first premise is that studies suggesting a higher degree of plasticity among certain types of adult stem cells 'are still open to multiple interpretations or require replication'. [10] The confused debate evoked by the publication of two recent studies in *Nature* - notably published only in the form of *letters* - on spontaneous fusion of stem cells confirms the soundness of this precaution. [11] The authors behind the letters do not - as reported by several news agencies and reiterated in *Nature Biotechnology* and in *British Medical Journal* [12]- call into question the possibility of a higher plasticity in adult stem cells than previously believed, nor do they deny the existence *in vivo* of the phenomenon of *proper* transdifferentiation in such cells. Instead, they draw attention to the phenomenon of spontaneous fusion taking place *in vitro* between embryonic stem cells and adult stem cells and suggest that differentiation properties in adult stem cells may be explained along this alternative pathway. As admitted by the group of researchers behind one of the letters, whether and to what extent cell fusion contributes to the formation of 'apparently transdifferentiated cells *in vivo* is currently pure speculation; however', the letter continues, 'our data raise a warning to the overzealous trend in stem cell research to conclude transdifferentiation or dedifferentiation of cells without careful examination of genotypes' [13].

In a subsequent Commentary in *Nature Biotechnology* to these studies [14], Ron McKay writes that although there are other studies indicating that cell fusion is not necessary for transdifferentiation to take place [15], experiments that 'directly show that stem cells are changing their fate' would still be needed to settle the question. And in a second Commentary on somatic stem cell plasticity published in the same issue of *Nature*

Biotechnology, Ihor Lemiscka pinpoints quite eloquently the current lack of experimental clarity on the issue of somatic stem cell plasticity:
'Recently, several publications have appeared that highlight the sometimes unexpected complications that can arise in studies of stem cell plasticity [16]. In short, these studies support the notion that extraordinary claims will require an extraordinary degree of experimental rigor' [17].

The second empirical premise on which the House of Lords Select Committee rely is oral and written evidence from a wide range of individuals (experts and lay people) and professional organisations. The result of these consultations was that very few experts - notably none of the consulted experts on *adult* stem cells! [18]- supported the view that the recent developments on adult stem cells had made embryonic stem cell research redundant. Positively formulated, the message conveyed to the Committee was that in the present situation of substantial ignorance about what represents the best route, adult stem cells and embryonic stem cells should not be viewed as research *alternatives*, but as 'complementary pathways to therapy' [19].

A third empirical premise relates to the processes of de- and re-differentiation which it would, according to the Committee, be unrealistic to understand thoroughly without using pluri-potent stem cells that are fully undifferentiated, i.e. embryonic stem cells. Although most of these studies can be undertaken on embryonic stem cells derived from *animals*, the Committee states that comparison with human embryonic stem cells would be required prior to attempts at applying the results to develop therapies. Consequently, it seems that a certain amount of basic research on human embryonic stem cells would still be necessary.
To further substantiate this point, the Committee also points to the complications involved in using adult stem cells instead to study the mechanisms underlying these processes: [20]
'If scientists are to dedifferentiate adult stem cells to pluripotency, prior to redifferentiation into a new cell type for therapeutic purposes, they must know whether they have done this correctly and whether the process is safe. Differentiation involves 'marking' the genetic material in a number of ways. These 'markings' (including chemical changes to the DNA and the interaction of specific proteins with it) are 'remembered' during cell division. If an adult stem cell is to be dedifferentiated prior to redifferentiation for therapeutic purposes, these markings must be correctly erased'.

A final premise underlying the Committee's rejection of the redundancy argument against human embryonic stem cell research is that although it may be probable that future developments will make research on human embryonic stem cells unnecessary, the present lack of any reliable predictions about which route represents the best option 'suggests that avenues of research should not be closed off prematurely' [21].

To complicate the issue of prediction a bit further, suffice it here to refer to a newly published study in which an *in vitro* method for analysing the mechanisms underlying nuclear reprogramming is presented and where it is suggested that differentiated somatic cells may not only be used to investigate the mechanisms underlying the processes of de-, re- and trans-differentiation, but also to produce 'isogenic replacement cells for therapeutic applications' [22]. If this prediction comes true, then it may very well be that a third route to cell-therapy will also see the day. However, as pointed out by Western and Surani: [23]
'...the applicability of this technology in producing reprogrammed cell lines for therapeutic purposes remains undetermined. A large amount of additional data are required before such a system could be applied to the generation of stem cells to be used directly in cell replacement therapies for human patients'.

To sum up the analysis so far: In its attempt at scientifically defending the case of embryonic stem cell research, the House of Lords Select Committee makes use of a combined strategy of

a) identifying a set of *basic* research challenges that need to be addressed prior to any attempts at therapeutic research applications, and
b) paying particular attention to recent developments in *adult* stem cell research.

Although hailing the therapeutic potential of adult stem cell research as 'great', the Committee concludes that in the meantime human embryonic stem cell research remains a 'clear' and 'strong' case, scientifically as well as medically [24]. This way of resolving the story of redundancy stands in contrast to Søren Holm's challenging account, where the available scientific evidence is not found to be strong enough to warrant human embryonic stem cell research. According to Holm, as long as the scientific evidence available does not demonstrate with 'unequivocal certainty' that human embryonic research is *necessary*, stem cell research should be restricted to 'non-destructive lines of research' [25]. Holm bases his negative verdict on two empirical premises: The lack of compelling evidence to suggest that:

i) human embryonic stem cell research will bring about therapeutic results 'much faster' than adult stem cell research, or
ii) that human embryonic stem cell research represents the only available pathway to generate cures for certain diseases [26].

By restricting himself to these two 'therapeutic' premises, Holm manages to build up a seemingly convincing case against destructive stem cell research on human embryos. The problem, however, is that there is one part of the story that is missing in his account: the part which, at least for the time being, represents the main chapter - *basic* scientific research. Holm rightly draws attention to a tendency among proponents of embryonic stem cell research of therapeutically *overselling* their case. However, by leaving out the premise of basic research in his *own* account, Holm risks ending up *rejecting too much too quickly*, and notably by employing the same contestable strategy as the enthusiasts of human embryonic stem cell research: the overselling - or perhaps a more appropriate wording in this case would be the 'over-killing' - of *therapeutic* arguments! A preliminary conclusion that might be drawn from this analysis therefore is: At present the use of *therapeutic* arguments, either to justify human embryonic stem cell research or to contest such research, should be rejected as *overuse* - if not necessarily as *abuse* - of scientifically available evidence. On the other hand, further analysis is needed to clarify whether the House of Lords Select Committee is justified in claiming that at least some of the *basic* research challenges identified in the *Report* cannot be adequately explored without allowing the use of (spare or created) human embryos.

Different Stories of Differentiation

This brings me back to the third empirical premise in the Select Committee's defence of human embryonic stem cell research: the need to understand and technically cope with the different processes of cell differentiation (research challenge 5 and 7 identified in the House of Lords *Report*). As stated by the Committee, in the first place only research on ES-cells derived from *animals* needs to be carried out. However, in order to be able to translate those results into something that can be used to develop therapies in the *human* sphere, similar studies on human embryonic stem cells to confirm - and eventually adjust the results - will also be necessary. Thus, the Committee argues, research into human embryonic stem cells is required whatever cell-type (ES-cells or adult stem cells) will be used in the future for therapeutic purposes, as 'apart from CNR, ES cells provide the only realistic means at present of studying the mechanisms and control of the processes of differentiation and dedifferentiation' [27].

It might however be that this is no longer the case, taking into account the newly published study of an *in vitro* method using differentiated somatic cells for investigating the mechanisms underlying these processes, but further research would be needed before the true potential of this method is clearly demonstrated. Besides, in case this method really proves to be efficient in deciphering the different processes of cell-differentiation, human embryonic stem cells would probably still be required as 'control-cells' to verify whether adult stem cells are really undergoing *complete* and *proper* dedifferentiation. Consequently, a certain amount of basic research on human embryonic stem cells seems to be unavoidable. A second argument in favour of including ES-cells in the study of the processes of de- and re-differentiation is that the sooner one gets a clear understanding of these processes, the sooner will it also be possible to start researching the therapeutic potential of *adult* stem cells. Thus, by including ES-cells as proper objects of study to understand the processes of re- and dedifferentiation, there are strong reasons to believe that valuable scientific results would be produced much faster than if basic research is confined to adult stem cells.

Competing Stories of Embryonic Origin

A remaining set of empirical premises that needs to be addressed relates to the question of embryonic *origin*, i.e. to the question which type of embryo would be preferable to use if human embryonic stem cell research is deemed necessary. In its *Report* the Select Committee suggests that *surplus* embryos from IVF treatment should be *first* choice, whereas the intentional creation for research purposes of human embryos by IVF should not be allowed 'unless there is a demonstrable and exceptional need which cannot be met by the use of surplus embryos' [28]. Two empirical premises are in operation behind the Committee's assessment: The huge amount of surplus IVF human embryos having been donated for IVF-related research (53497) and the small number of additional IVF embryos having been found necessary to create for that specific purpose (118) in the period between August 1st, 1990 (when the Human Fertilisation and Embryology Act came into force) and March 31st, 1999 [29]. As for the possibility of using CNR-technology to produce embryos for research, the Committee argues from the premise that CNR represents a *powerful* research tool to understand the mechanisms underlying the processes of de- and re-differentiation. However, in spite of its positive verdict, the Committee suggests that production of CNR-embryos should be restricted in the same way as is the creation of research-embryos by IVF, in order to minimise the risk of *instrumentalisation* of human life to a degree beyond that of a research practice restricted to the use of surplus IVF embryos [30]. Finally, the Committee refers to the 'excellent record' of the Human Fertilisation and Embryology Authority (HFEA) as convincing evidence for the possibility of monitoring embryonic research in a way sufficient to protect against any unwarranted practice of embryonic research, reproductive cloning included [31].

Concluding Remarks about an Unfinished Story

The devil is in the details. This well-known remark seems appropriate to sum up the results of the stem cell controversy so far. As the above analysis has shown, the abundant use of *therapeutic* arguments to justify human embryonic stem cell research lacks empirical foundation. It represents unwarranted exploitation of scientifically available evidence. This negative conclusion also seems to hold true for the opposite case, i.e. of using therapeutic arguments to question the necessity of human embryonic stem cell research. Although failing to get all the details right does not necessarily mean that the overall story will end up distorted, in the case of stem cell research it seems justified to say that empirical details deserve more attention in the ethical debate than hitherto given. Consequently, the empirical case in favour of a limited amount of destructive *basic* research on human embryos should also be included in the account. As to the question of embryonic origin, the huge amount of surplus embryos from IVF supports their selection as *primary* sources for basic stem cell research. On the other hand, available evidence also suggests that the use of CNR - and thereby the production of human embryos - might become necessary in order to understand the mechanisms underlying the processes of de- and re-differentiation. Third, there is little evidence available to support the view that creation of IVF-embryos for stem cell research is necessary. Finally, the recent publication of a study reporting the discovery of a previously unknown type of cells in *adult* bone marrow - so-called multipotent adult progenitor cells (MAPCs) - with an *in vitro* ability of differentiation comparable to human embryonic stem cells and with the capacity of organ-directed differentiation *in vivo*, [32] makes clear that this is an ever-evolving narrative.

*Forthcoming in Cambridge Quarterly of Health Care Ethics Vol. 12, No. 4, Fall 2003

References

The author thanks the EMPIRE project, funded by the European Commission, DG-Research for stimulus and support.

[1] Holm, S, 'The Ethical Case Against Stem Cell Research', forthcoming in *Cambridge Quarterly of Health Care Ethics*
[2] Holm, S, 'Going to the Roots of the Stem Cell Controversy', forthcoming in *Bioethics*
[3] See note 2, Holm, S, (2002), p.15
[4] See note 2, Holm, S, (2002), p.2
[5] House of Lords. Stem Cell Research. Report from The Select Committee. February 13th 2002, 2.12 (hereafter House of Lords *Report*)
[6] See note 1, Holm, S, (2002) p.4. See also note 2, Holm, S, (2002), p.11
[7] The European Group on Ethics in Science and new Technologies to the European Commission, 'Ethical Aspects of Human Stem Cell Research and Use', November 14th 2000: 2.5
[8] See for example, Hansen, J-ES, (2002), Embryonic Stem Cell Production Through Therapeutic Cloning Has Fewer Ethical Problems than Stem Cell Harvest from Surplus IVF Embryos, *Journal of Medical Ethics* 28, pp.86-88
[9] See note 5, House of Lords *Report* 2002: 2.11-2.12
[10] See note 5, House of Lords *Report* 2002: 3.15
[11] Terada, N, Hamazaki, T, Oka, M, Hoki, M, Mastalerz, DM, Nakano, Y, Meyer, EM, Morel, L, Petersen, BE, Scott, EW, (2002), 'Bone Marrow Cells Adopt the Phenotype of Other Cells by Spontaneous Cell Fusion', *Nature* 2002; 416 p.542-545; and Ying, Q-L, Nichols, J, Evans, EP, Smith, AG, (2002), 'Changing Potency by Spontaneous Fusion', *Nature* 416 p.545-54
[12] For example, 'Flaws in Studies of Adult Stem Cells Dash Medical Hopes', *Daily Telegraph* March 14th 2002; 'Study Weakens Anti-Abortionists' Adult Tissue Claim', *The Independent* March 14th 2002; 'Experts Question Studies Suggesting Adult Stem Cells Won't Work' *CNSNews.com* March 15th 2002; 'Turning Back the Clock', *Nature Biotechnology*. May 20th 2002: 411: '...the finding that ES cells can fuse with adult stem cells in coculture throws into question whether previous observations of transdifferentiation were due to reversion of adult stem cells or expansion of abnormal hybrids'; Mayor S, 'Adult Stem Cells May Not Be Able to Differentiate Into Other Cell Types', News roundup *BMJ* March 23 2002;324:696. See also, 'Researcher Says Media Distorted Adult Stem Cell Studies', *CNSNews.com* March 25th 2002, where, Dr. Naohiro Terada, one of the researchers behind one of the studies on cell fusion is quoted as saying: '..our message was somehow distorted by the media', and '...we never said adult stem cells are no longer hopeful, nor dangerous. If someone took our message that way, that is a misinterpretation'.
[13] See note 11, Terada et al. 2002: 544

[14] McKay, R, (2002), 'A More Astonishing Hypothesis', *Nature Biotechnology* 20, p.426
[15] McKay's reference is to the following studies: Rolink, AG, Nutt, SL, Melchers, F, Busslinger, M, (1999), 'Long-Term *In Vivo* Reconstitution of T-cell Development by Pax5-Deficient B-Cell Progenitors', *Nature* 401 pp.603-606; Seale P et al, (2000), 'Pax7 Is Required for the Specification of Myogenic Satellite Cells' *Cell* 102, pp.777-786; McKinney-Freeman, SL, et al, (2002), 'Muscle-Derived Hematopietic Stem Cells Are Hematopoietic in Origin', *Proc. Natl. Acad. Sci. USA*, 99 pp.1341-1346
[16] In addition to the two studies on cell fusion already referred to in note 11 and McKinney-Freeman et al's study referred to in the previous note, the author also mentions the following study: Morshead, CM, Beveniste, P, Iscove, NN, van Der Kooy, D, (2002), 'Hematopoietic Competence is a Rare Property of Neural Stem Cells that May Depend on Genetic and Epigenetic Alterations', *Nature Medicine* 8, pp.268-273
[17] Lemischka, I, (2002), 'Rethinking Somatic Stem Cell Plasticity' *Nature Biotechnology*, 20 p.425.
[18] The Committee received replies from four internationally renowned experts on adult stem cells: Professor Helen Blau, Stanford University School of Medicine, Dr. Jonas Frisen, Karolinska Institute, Stockholm, Professor Nadia Rosenthal, The European Molecular Biology Laboratory, Monterotondo-Scalo and Professor and Director Angelo Vescovi, The Stem Cell Research Institute, Milan.
[19] See note 5, House of Lords *Report* 2002, 3.16
[20] See note 5, House of Lords *Report* 2002, 3.18
[21] See note 5, House of Lords *Report* 2002, 3.19-3.21.
[22] Håkelien, A-M, Landsverk, HB, Robi, JM, Skålhegg, BS, Collas, P, (2002), 'Reprogramming Fibroblasts to Express T-cell Functions Using Cell Extracts', *Nature Biotechnology* May 20, pp.460-466
[23] Western, PS, Surani, MA, (2002), 'Nuclear Reprogramming - Alchemy or Analysis? A New Strategy for Nuclear Reprogramming Using Cell Extracts Induces Fibroblasts to Express Hematopoietic and Neuronal Responses, *Nature Biotechnology* May 20, p.445-446
[2] See note 5, House of Lords *Report* 2002, 3.22 (f). For the wording see also the *Report's* Summary of Conclusions and Recommendations, paragraph 4.
[25] See note 2, Holm, S, (2002), p.15-16. See also note 1, Holm, S, (2002), p.6
[26] See note 1, Holm, S, (2002), p.6
[27] See note 5, House of Lords *Report* 2002, 3.17
[28] See note 5, House of Lords *Report* 2002, 4.28
[29] See note 5, House of Lords *Report* 2002, 4.26. Reference is made to the *Human Fertilisation and Embryology Authority* (HFEA), Ninth Annual Report and Accounts, 2000
[30] See note 5, House of Lords *Report* 2002, 5.14
[31] See note 5, House of Lords *Report* 2002, 5.24
[32] Jiang, Y, Balkrishna, N, Reinhardt, RL, Schwartz, RE, Keene, CD, Ortiz-Gonzalez, XR, Reyes, M, Lenvik, T, Lund, T, Blackstad, M, Du, J, Aldrich, S, Lisberg, A, Low, WC, Largaespada, DA & Verfaille, CM, (2002), 'Pluripotency of Mesenchymal Stem Cells Derived from Adult Marrow', *Nature* (advance online publication) June 20th, 2002: www.nature.com, doi: 10.1038/nature00870.

Carol Gilligan and her Different Voice

Heta Aleksandra GYLLING
Department of Moral and Social Philosophy
University of Helsinki

Historical Background

The most important questions of normative moral and political philosophy have been and still are 'How should individuals behave?' and 'How should life in human societies be organised?' Philosophers have seldom been interested in empirical findings concerning people's actual answers to these questions or in their beliefs about what they think would be or should be their view on the subject. And, of course, people tend to give to these questions the kind of answers that tell others what a nice and decent lot they are. Hence, it may be difficult to tell whether the answers reflect anything other than their own expectations.

The interest in moral sociology and empirical research has – outside the domain of cultural and social anthropology – faded and works of this kind have become mere historical curiosities, most often related to in Edward Westermarck's work [1]. For him ethics was a sociological and psychological discipline and he forcefully denied the existence of general moral truths and the objective validity of moral judgements or universal truths. His own interest lay mainly in the study of moral consciousness and the existence of cultural differences, but not in an attempt to establish and justify general or universal rules for conduct. But neither were others after him interested in normative questions so that very few academic philosophers had satisfactory answers to these aforementioned questions. Ethical theories were drawn towards subjectivism, relativism and nihilism, while political thought seemed to condone totalitarian thinking without qualms.

Finally, after the Second World War, the interest of many philosophers shifted from the then predominant linguistic analysis to the consideration of the function and role of such entities and principles as duties, rights, liberty, fairness and justice in moral and political life, thus re-enforcing the liberalist tradition, both in its economic and political forms. And a new discussion started when John Rawls published his *Theory of Justice*, claiming that the correct set of principles which can be accepted by any free and rational individual (which equates to any rational decision-maker) will be found behind the veil of ignorance. These allegedly unbiased decision-makers are not only well acquainted with the functioning of the human psyche but also able to accurately predict various effects that different social arrangements may have on individual citizens, their lives and sense of justice.

The principles of justice that Rawls put forward can be deduced directly from the imaginary setting he calls the original position, which includes assumptions concerning general knowledge and lack of personal knowledge, risk-taking, envy and the tendency of individuals to promote their own welfare. For Rawls, justice was a set of principles, which is designed to regulate the attribution of the rights and duties of individuals and the distribution of social, political and economic benefits and burdens in a society.

Understandably this interpretation of a just society did not appeal to all, whether proponents of liberalism or not. Some rejected the Rawlsian view because of its deontological nature, favouring instead Millian consequentialist-utilitarian formulations, and some, notably Robert Nozick, claimed that Rawls went wrong at the very outset of his theory. For Nozick, justice is not a pattern to be aimed at but a process which is founded on pre-existing rights and which defines what individuals are historically entitled to. The state in Nozick's model was a modest concoction, and its main function is to safeguard the historical entitlements of its citizens against active violence, deception and theft. For him, every attempt to redistribute property constitutes a violation of the rights of individuals, in the name of procedural justice.

But the main critique wasn't aimed just at Rawls but against liberalism and its inherent individualism. The so-called communitarians did not accept the core idea of abstractedly rational individuals who can freely and without knowing their position, choose a good set of principles for the just regulation of all aspects of social life. They resent the doctrine that the state should be value neutral and that the public authorities should condone all the different conceptions of good that people may have. As an alternative, most of them advocate, in one form or another, the view that communities not only play, but ought to play, a decisive role in the definition and formulation of people's identities, moral values and beliefs concerning justice.

But the critique of liberalism did not end there. Even if most of the so-called original communitarians thought it enough to emphasise the value of one's history, social role and traditions, some women wanted to take the critique a bit further and question the basic and dominant concept itself, namely justice. First, some came to believe that communitarian types of thinking should surpass liberal principles because women as mothers and carers could become more influential in a world not exclusively governed by the supposedly male rules of public policy and finance. This belief is legitimate in the limited sense that caring and mothering are as closely knitted to the fabric of society as politics and commerce, and should therefore be given their due when public affairs are planned and regulated. But it is more questionable when it comes to the positions women could actually achieve in the kinds of communities referred to by, for instance, Taylor and MacIntyre.

What really changed the whole picture was the acquittal of the universalism of Enlightenment liberalism and its replacement with ideas, strongly influenced by Jean-Jacques Rousseau and Immanuel Kant (especially Rousseau, being, as he was, a representative of Romanticism rather than of Enlightenment). Rousseau's influence lay in his views concerning reason and emotion, and Kant's not least in his views on questions of individual freedom and active citizenship, which in Kant's thinking was quite beyond female capacities. In Romanticism people were primarily seen as members of their historical nations and cultures, not as separate beings with their own individual wills and desires. And in Rousseau's case an additional spice was his firm belief in the essential difference between men and women, who – as weak and vulnerable but moral creatures - cannot base their life on autonomous choice and rational thinking but rather have to rely on feeling and emotion, thereby asserting their particular nature.

The reasons why some feminists have sought inspiration and confirmation for their own feelings and convictions from Rousseau may of course vary. Some may appreciate Rousseau's conceptions on the interplay of feeling, sensibility, sense of community and social influence and think that his ideas on individual autonomy and maybe even the harmfulness of arts and sciences should be taken seriously in our time. And some may sincerely believe that there actually is a positive difference between men and women, independent of education and treatment as a child; the difference not just being a particularistic emotional nature created by the historical fact that women were denied the opportunity to develop their rational capacities – as, for instance, Mary Wollstonecraft claimed. Different interpretations can be found depending on whether one tends to feel that Rousseau devalues or overrates women and their ability to be at the same time both reasonable and moral creatures. The final interpretation may also depend on whether one happens to be an ardent admirer of Freud or not:

'I cannot evade the notion (though I hesitate to give it expression) that for women the level of what is ethically normal is different from what it is in man. Their superego is never so inexorable, so impersonal, so independent of its emotional origins as we require it to be in men. Character-traits which critics of every epoch have brought up against women – that they show less sense of justice than men, that they are less ready to submit to the great exigencies of life, that they are more often influenced in their judgements by feelings of affection or hostility – all these would be amply accounted for by the modification in the formation of their super-ego which we have inferred above.'[2]

From Cold Justice to Loving Care

As such, love, friendship and care are good and valuable and the reintroduction of these concepts into ethical discussion cannot but be praised. But what can and should be criticised is how they have been introduced. Many feminists seem to believe – and this is partly due to Carol Gilligan's alleged findings – that caring ethics or ethics of care is something inherently suitable for women, or even that it embodies an innate moral virtue, which cannot be possessed by ordinary mortal men. It may be that those who say that human beings are first of all members of their societies and communities, occupiers of their socially and culturally determined roles, and moral agents whose ethical values are defined by the linguistic and historical context, are quite right. But what they tend to ignore is that cultures and societies are prone to change and that even if two societies seem to belong to an almost identical historical tradition, sharing basic cultural patterns and settings, whilst it may be possible for their members not only to understand each other but also to communicate without excessive misunderstandings, their inner moral and social texture may be almost incompatible.

People belonging to the Western tradition in a sense belong to a shared historical framework which unites them (mainly when confronted by non-western thinking). We share certain patterns in our ways of thinking about humanity, individual rights and concepts of time but otherwise we are not divisible into pre-established groups, nations and cultures but rather represent a mixture of different traditions and ways of thinking, sometimes feeling united by national boundaries, sometimes not. Most importantly, our social and political traditions and ideals vary; these variations strongly influencing the way people think and conceptualise the world. Our views on what we are, how we see ourselves and our social position in a cultural context is affected by the structure of the society, its educational system and, for instance, how family roles are seen. Also, what must be kept in mind is that the way we express ourselves and how we conceptualise our social web in moral terms may depend on how the language presents itself to its speakers.

Carol Gilligan's Study and its Background

Carol Gilligan's well-known work on the distinction between an ethic of care and an ethic of justice was based on interviews and stems from Lawrence Kohlberg's model of moral development. Kohlberg's idea was that there are six different stages in moral development. In order to analyse these levels he created a set of ethical dilemmas designed to show how different individuals climb from lower stages of morality – like utilitarianism - to higher morality, meaning Kantian moral reasoning. Unfortunately, he neglects to explain why

Kantian ethical thinking should be seen as the peak of sophistication in moral reasoning. Also, he has been accused of being unable to distinguish between different theories concerning the epistemological ground of ethical judgements and of mixing up the rightness of moral acts with the motivation behind moral action [3].

Gilligan's hypothesis was that the fact that Kohlberg's study showed girls ranking lower in their moral development than boys, had to have some other explanation than the inadequacy of girls' morality. After her interviews with both men and women, Gilligan was ready to claim that the lower ranking was due to Kohlberg's inability to see how women and men think differently about ethics. From the time that they are young, women learn to think about ethics in terms of care, responsibility and interdependence in relationships while men rely on ethics of justice, making calculations and balancing people's rights, the justice of their claims and fair play. But exactly why and how this different voice develops, whether it marks an essential difference between women and men and, most importantly, whether this difference should have normative value, remains unclear. In what follows I try to point out some of the problems that Gilligan's research either contains or has created. It has been and still is influential and in my view has led quite a lot of researchers, both philosophers and others, astray.

The Nature of the Empirical Research [4]

The study consists of three partly separate and partly interrelated interviews about conceptions of self and morality and individual experiences of conflict and choice, complemented with further questions in order to ascertain that the logic of the interviewee's thought wasn't misunderstood. The first of them, the college student study, 'explored identity and moral development in the early adult years by relating the view of self and thinking about morality to experiences of moral conflict and the making of life choices' [5]. Twenty-five senior students, selected at random from those who were to take a course on moral and political choice, were interviewed at the time and again five years after their graduation. The second one, the abortion decision study, explored women's feelings and reactions – or at least what they thought they should feel about abortion. For this study, twenty-nine women were chosen, ranging in age from fifteen to thirty-three and, according to Gilligan, representing different ethnic backgrounds and social classes. Some of them were single, some married, a few had a young child. They were interviewed during the first trimester of a confirmed pregnancy at a time when they were considering abortion. And finally, the third one was 'generated by these studies concerning different modes of thinking about morality and their relation to different views of self' [6] which were further analysed in the rights and responsibilities study, involving a sample of males and females matched for age, intelligence, education, occupation, and social class at nine points across the life cycle, the age groups being 6-9, 11, 15, 19, 22, 25-27, 35, 45, 60, with a total sample of 144. There was also a sub-sample of 36 individuals who were interviewed more intensively. The questions concentrated on conceptions of self and morality, experiences of moral conflict and choice, and judgements of hypothetical ethical dilemmas [7].

The first obvious question concerns the scientific validity of the research. Is it reasonable to believe that the number of interviewees involved gives an adequate picture of what and how women think about morality in comparison to men? Gilligan says in the introduction of her book that:
'the different voice I describe is characterised not by gender but theme. Its association with women is an empirical observation, and it is primarily through women's voices that I trace its development. But this association is not absolute, and the contrasts between male and female voices are presented here to highlight a distinction between two modes of thought

and to focus a problem of interpretation rather than to represent a generalisation about either sex'. [8]

It may be that Gilligan herself didn't want to claim an overall universalisability to the answers, but others referring to her results have tried to do so without worrying whether the sample was really reliable and representative, that is, whether it reflects in any serious sense what men and women, in opposition to each other think and say about morality. And – most importantly - whether it is possible to deduce anything reliable about people's moral beliefs from what they *say* they think. We may ask whether this different voice really exists in the sense that it correlates with either sex or gender, instead of being something both men and women employ in certain moral contexts. Also, it is possible to explain the results in a less gender specific way. For instance Susan Okin has remarked that powerless groups often learn empathy because they themselves need the protection of others and:
'as subordinates in a male-dominated society, [women] are required to develop psychological characteristics that please the dominant group and fulfil its needs'.[9]

Still, it is advisable to keep in mind that if it is women's subordinated role that has historically made them more care-oriented, this does not in itself prove that women – as female human beings – would naturally be the care-oriented ones. Okin's suggestion only means that those who are subordinates in a certain society may have to develop certain psychological characteristics – a thing that may have happened to most women (not necessarily to all women but to those who happen to speak in terms of caring ethics).

Defending the scientific validity of her research, Gilligan emphasises the fact that the interviewees had different social and cultural backgrounds. Unfortunately, this claim in itself proves nothing. Even if ethnic and social differences may be reflected in their opinions and views - without affecting the shared female voice - the fact remains that the interviewees were all products of the American society and, as children immersed in a pool of social and political ideas and ideals, are hardly identical with European counterparts, who may exemplify diverse ways of constructing the social and political reality. When being interviewed, people tend to take their own reality for granted and don't necessarily analyse their choices in the light of unfamiliar societies, unaccustomed social values and different legal frameworks. For instance, when asked about abortion, some women seemed to feel that they had to choose between child and job. Not in the sense that they would have pondered whether having a child at the time might seriously hamper their career but in the sense that they found themselves in a position in which both keeping their job and having a child appeared to be a non-existent option. While talking about their responsibilities towards the potential child, they didn't question the responsibilities of the state and political order, apparently forcing women to think about abortion in these terms – instead of providing a child-care system so that women would have a real choice.

I believe that women's tendency to talk in terms of care ethics might be less pronounced in countries where nobody at the age of leaving school would seriously consider the option of becoming a housewife as a life-long career. I am not saying that nobody in Finland, for instance, would dream of that but rather that ethics of care may taste a bit different in a culture where the wish not to work outside home always needs an explanation – unlike in America. On the other hand, I would also be surprised if men, after their parental leave, would find the concepts of caring and responsibility totally alien to them.

Personal and intimate experiences affect us; hence the discovery that women considering abortion don't talk about it in terms of rights and justice is hardly surprising and doesn't prove that women would always think about termination of pregnancy in terms of care and responsibility. The situation could be seen as somewhat analogous to what David Hume said about justice and human nature. He was convinced that one of the reasons why we need the concept of justice is our inability to extend altruism to all our fellow-beings. A good and functioning family does not need any principles of justice to guide it when forced to share insufficient daily bread: love and mutual concern take care of the division. But in meagre times, without rules and principles, a group of strangers would not last long.

Gilligan is most certainly right when stating that "women's place in man's life cycle has been that of nurturer, caretaker, and helpmate", but this partly historical, partly contemporary fact does not mean that all women would – and definitely not that they should – "define themselves in a context of human relationship but also judge themselves in terms of their ability to care" [10]. Much the same can be said about the following quotation:

'To understand how the tension between responsibilities and rights sustains the dialectic of human development is to see the integrity of two disparate modes of experience that are in the end connected. While an ethic of justice proceeds from the premise of equality – that everyone should be treated the same – an ethic of care rests on the premise of non-violence – that no one should be hurt. ... This dialogue between fairness and care not only provides a better understanding of relations between the sexes but also gives rise to a more comprehensive portrayal of adult work and family relationships.'[11]

This could be true if we had good reasons to believe women really think and talk only or mostly in terms of ethics of care. The women Gilligan talked to seemed to do so, but that in itself doesn't prove a thing about women and men in general.

Still, these quotations make me wonder whether we are really supposed to think that a clear-cut division between rights and justice on one hand and care and responsibility on the other can be made. Are they incompatible in a sense that proponents of the former could not possibly value caring, spontaneous human relationships and responsibility? Or is ethics of care just one aspect in our ethical cogitation? Is it merely one moral perspective, particularly suitable and applicable in the realm of family and friendship? Or is care ethics, as at least some feminists seem to think – superior to all male engendered ethical theories and extendable to public affairs too? In order to be able to answer these questions, I must try to analyse my interpretation of Gilligan's interpretation of the difference between these two ethical approaches.

Ethics of care seems to require certain moral capacities, a proper mental state and a disposition to see others not only as moral agents governed by abstract principles. People have to feel responsibility for others, they have to be able to feel sympathetic towards suffering and they need an impulse to react with caring and sympathy. Seemingly, they are supposed to see moral dilemmas and conflicts as a sympathetic helper whose main aim is to solve the problems in an amiable ambience. Like Socrates, one should help others to see the obvious solution. But is ethics of justice void of these moral capacities? Many have noted [12] that applying abstract ethical principles requires moral capacities and characteristics, and the ability to interpret human life and behaviour. Without a sense of justice, without sympathy and imagination we cannot even recognise ethical problems and their exact nature, which is a precondition for analysing and solving them.

Ethics of care is seen as warm and close in contrast to cold and distant justice. According to Lawrence Blum:

'For Gilligan, morality is founded in a sense of concrete connection and direct response between persons, a direct sense of connection which exists prior to moral beliefs about what is right or wrong or which principles to accept.' [13]

And also:

'For Gilligan, care morality is about the particular agent's caring for and about the particular friend or child with whom she has come to have this particular relationship. Morality is not (only) about how the impersonal "one" is meant to act toward the impersonal "other". In regard to its emphasis on the radically situated self, Gilligan's view is akin to Alasdair MacIntyre's (After Virtue) and Michael Sandal's (Liberalism and the Limits of Justice)'. [14]

But talk about justice doesn't necessarily and primarily imply reference to moral or legal rights. Rights-based talk and thinking is typically and traditionally an American phenomenon, not shared by us all. We do have different justice oriented moral theories and also different everyday parlances. Gilligan may give an accurate description of American male moral language when picturing American men as moral agents who, when talking about morality, emphasise the importance of rights. This, however, couldn't be said of, for instance, a Finnish speaker. This difference is best illustrated by how the use of verbal expressions may reflect the way we construct morality. Let us suppose that person A has some vital information which will affect me and my life. After finding this out, how would I express my wish to be enlightened? I presume that were I an American, I would immediately inform A of my allegedly inalienable right to know. But, alas, being a Finn I would never even think of formulating my demand in terms of rights but instead would remind the person in question of her duty to tell, that is, in everyday language I would cast an accusing eye on her and say: "You have to tell me!" In our culture duties have traditionally formed the core of moral discussion and it is only during the past years – and under American influence - that the idea of emphasising individual rights has entered the Finnish existence.

How to Interpret It?

The presupposition throughout the study clearly is that since women's social role is amidst the family and their activities have traditionally been confined to private matters, (like being the natural bearers of children and their moral educators) unlike men, (who mainly operate on social and political levels which are based on voluntary, human-made associations, based on covenants, contracts, blind justice and impersonal rights), it is only appropriate that women should concentrate on the individual and psychological aspects of moral problems. And one could well conclude from Gilligan's results that women should stick to their traditional roles and not try to invade male domains. Whether it is really well founded to talk about women as a group in connection of values and morality, basing the whole lot on three minor studies, is another question. I myself find it preposterous to assume that there is something morally relevant, which all women share, no matter what age and culture they represent. This assumption isn't of course even theoretically possible if we are ready to reject the vulgar forms of essentialism, namely the belief in morally relevant biological determinants. Women don't constitute a morally separable group from men and they don't form a group of their own in the sense that one could claim that the values binding women emerge from shared womanhood and therefore differ from those befitting men.

Carol Gilligan's work has created quite a lot of discussion about morality; its nature and purposes. Some want to see it as a fruitful addition to traditional moral philosophy, shedding (empirical?) light on how at least some women tend to think about morality and how it differs from the traditional (allegedly) justice-oriented male voice. And some, like Sara Ruddick [15], have taken Gilligan's ideas even further, claiming that caring and what she calls maternal thinking should be extended to international politics, agreeing with Ned

Noddings [16] that abstract action-guiding principles are not enough if proper morality is to be achieved. The practice of mothering may be - as Ruddick suggests – what:
'makes reflective feeling one of the most difficult attainments of reason. In protective work, feeling, thinking, and action are conceptually linked; feelings demand reflection, which is in turn tested by the feelings it provokes. Thoughtful feeling, passionate thought, and protective acts together test, even as they reveal, the effectiveness of preservative love' [17].

But whether this is what morality is supposed to be remains questionable. Some do agree with Ruddick and find this preservative love a suitable approach to be applied even in public affairs and international politics. Personally, I think that mothering, viz. treating others like children sounds more like colonial politics than democratic co-operation. And even if we were to confine ethics of care to domestic questions, we may still ask whether we all agree on what mothering might be, or do we allow the existence of different kinds of mothers? Many women of colour have felt, especially in the U.S, that privileged white women have for years tried to speak for all American women. If this is true, then it is probably also true that they try to speak for us Europeans as well.

References

[1] Westermarck, Edward, (1862-1939). Main works include The History of Human Marriage (1891), The Origin and Development of the Moral Ideas, 2 vol. (1906-08) and Ethical Relativity (1932).
[2] Freud, Sigmund, (1961), Some Physical Consequences of the Anatomical Distinction between the Sexes. In Strachey, J (Ed.), *The Standard Edition of the Complete Works of Sigmund Freud.* Vol. 19. London: Hogarth Press
[3] Green, Karen, (1995), *The Woman of Reason. Feminism, Humanism and Political Thought.* Cambridge: Polity Press, p.153
[4] Gilligan, Carol, (1982), *In A Different Voice*, USA, Harvard University Press
[5] Gilligan, Carol, (1982), ibid, p.2
[6] Gilligan, Carol, (1982), ibid, p.3
[7] Gilligan, Carol, (1982), ibid, p.3
[8] Gilligan, Carol, (1982), ibid, p.2
[9] Okin, Susan, (1990), 'Thinking like a Woman', in Rhode, D, (Ed.), *Theoretical Perspectives on Sexual Difference*. New Haven: Yale University Press, p.154
[10] Gilligan, Carol, (1982), op cit p.7
[11] Gilligan, Carol, (1982), op cit p.174
[12] Among others, Kymlicka, Will, (1990), *Contemporary Political Philosophy*, Oxford: OUP, p.266
[13] Blum, Lawrence, (1988), 'Gilligan and Kohlberg: Implications for Moral Theory', *Ethics*, Vol. 98 p.476
[14] Blum, Lawrence, (1988), ibid, p.474
[15] Ruddick, Sara, (1989), *Maternal Thinking*, Boston: Beacon Press
[16] Noddings, Ned, (1984), *Caring: A Feminine Approach to Ethics and Moral Education*, Berkeley: University of California Press
[17] Ruddick, Sara, (1989), op cit, p.70

Who Should Decide and Why? The Futility of Philosophy, Sociology and Law in Institutionalised Bioethics, and the Unwarrantability of Ethics Committees

Tuija TAKALA
University of Helsinki, Finland
University of Manchester

We live in a world where bioethical issues need to be regulated. Our multicultural and pluralistic modern societies, and the vast advances in life sciences, create bioethical problems that have to be settled somehow. There is a practical need for someone to know what to do when ethical uncertainties arise. In this paper I will study the question of who should make decisions concerning bioethical issues and why. Within academic bioethics there is constant rivalry between the main disciplines involved. Within the disciplines themselves there also are rivalries between different theoretical and methodological views. This makes academic bioethics a poor contender in the competition for the decision-maker-in-bioethical-issues award. Many seem to think that this is why we need ethics committees, where various interest groups and experts are represented, to solve the issues that academic bioethics is unable to handle. In what follows I will delineate the limitations of academic bioethics and show that it is indeed, as it now stands, unable to provide us with ethical solutions on which we could justifiably base regulations. I will also argue that ethics committees, as they now stand, are not the answer either. There is no academic consensus, public opinion, or democratic process in place to justify their role as the proper guardians of ethics.

Ethics Experts

The term 'bioethics' can refer to a number of different phenomena. For some, bioethics is first and foremost a scientific enquiry, whilst for others it means what is good, right and true in medical and related issues. The general public usually understands the word in the latter sense, but they often expect bioethics to also function as scientific enquiry, to produce the right answers. Surprisingly many academics working in the field have also lured themselves into thinking that their epistemic expertise in bioethical matters gives them a position of moral authority when real-life decisions are made.

The notion of an 'ethical expert' is in many ways problematic. Being an expert in other fields of study includes at least the following elements. Firstly, that the person knows the theories and traditions of her field. Secondly, that she has familiarised herself with the contemporary issues of the field. Thirdly, that she is able to give 'right' answers to questions requiring her particular expertise. Right answers are those that are grounded in the current paradigm of her discipline. And fourthly, that she is institutionally accepted as an expert. That is, for instance, that she holds important positions of trust and has published in recognised journals. Bioethical expertise, especially when this means the ability to give answers to questions concerning right action, is problematic because there is no one bioethical paradigm that is (or should be) recognised by all. Answers to normative questions are always relative to the normative premises chosen, and they are only right within restricted frameworks. Further, it seems obvious to many that ethical expertise is something that you do not learn from books. And since the discipline is relatively young, and multidisciplinary, even the other criteria for expertise are difficult to meet.

The Role of Different Disciplines

When bioethics is understood as a scientific enquiry, it falls within the scope of many different disciplines, depending on the questions to which it is thought to seek answers. The two main types of question that become relevant here are the normative and the descriptive. If bioethics is supposed to tell us what we should do, it is closely connected to those academic subjects that claim to have normative answers to practical questions. Moral philosophy, is the paramount contemporary example, although certain kinds of theology, as well as legal positivism in jurisprudence, can also fit into this category. If, on the other hand, bioethical enquiries are expected to produce information about the feelings and attitudes related to medical and scientific practices, the most important disciplines are empirical psychology and sociology. An intermediate alternative is to interpret bioethics as a mix of all the disciplines that are interested in seeking knowledge about the ethical issues arising from medicine, health care, and the biosciences. In this case, normative and descriptive studies can be seen as equally good tools in bioethical research [1].

If we take the general opinion that bioethics should produce adequate normative responses to current ethical issues seriously, we may want to consider the many different methods that moral philosophy has to offer. The best known of these is 'principlism', a doctrine according to which we can solve bioethical problems by resorting to a few simple moral principles that are, supposedly, shared by all [2]. In the so-called 'engineering models' a classical theory of moral philosophy, such as virtue ethics, deontology or utilitarianism, is applied to bioethical problems directly and without much contextual consideration [3]. 'Casuistry' is a method in which some basic moral problems and their somehow inevitable solutions are first identified, and more complex questions are then settled by drawing analogies between the sets of cases [4]. Yet another method is to make use of philosophical analysis, both conceptual and argument-related, in the deconstruction of suggested solutions to moral problems and the construction of new ones [5]. This approach acknowledges that the proposed solutions are only valid when certain explicated presuppositions are accepted. Accordingly, normative conclusions in this framework take the following form: 'If you believe X and Y, then Z is the morally right path to follow'.

The weight of empirical data in these various approaches depends on the normative premises chosen. The keener the theory is on actual outcomes, the more important the empirical knowledge, and, conversely, the more absolute the moral rules, the less empirical evidence is needed. By and large, those with consequentialist convictions need to understand the nature and state of the social reality where moral problems arise, while those with strict deontological ideals can sometimes do without such knowledge.

In addition to philosophical and social-science doctrines, a number of religious approaches to bioethics compete for recognition as academic enterprises and normative forces. As in the case of other approaches, the conclusions reached within them are heavily dependent on the methods and normative premises chosen.

The Expected Prescriptivity

Those who think that normative moral statements cannot be produced by scientific enquiry often hold that academic bioethics can, nevertheless, provide facts and views concerning ethical matters. These matters can include the beliefs of the general public, the opinions and attitudes of health care professionals and scientists, the history of medical ethics, the legal situation in a given country or culture, or the words and concepts used in current bioethical discussions. It is often forgotten, however, that when bioethics is approached descriptively, the outcomes of the studies are also purely descriptive. Journalists and the general public frequently make the mistake of assuming that normative conclusions follow from empirical facts without further justification or analysis. What is more, even the academics themselves can read normative meanings into descriptive findings. Perhaps this is an indication of the deeply rooted underlying expectation that bioethics should, after all, be a normative enterprise.

In order to further knowledge and understanding regarding bioethical issues, all existing approaches seem to have a legitimate role to play. When it comes to the expectations laid on bioethical experts, however, all academic approaches are equally inadequate. A moral philosopher can only speak from within her own normative framework, and there are seldom proper arguments with which to convince those who have different views. The same can be said about theological ethicists and legal positivists – that is, those who hold that prevailing laws should be observed because they are prevailing laws. All normative approaches to bioethics produce normative conclusions, but none of them can convincingly claim superiority over all the alternative approaches. Social science studies produce competing interpretations of the world, and it would be biased to raise one interpretation above the others. And even when sociological surveys produce widely accepted facts, these do not, as such, help us to identify the right moral answers. Similar observations apply to anthropological, historical, legal and conceptual studies.

Real-life bioethical decision-making in the institutional form of ethics committee work is a type of political process that has very little to do with academic bioethics, although it can, to a certain extent, benefit from it. Ethics committees, whether advisory or legislative, bring together a group of various experts and lay people, and entrust them with the task of ensuring the ethical acceptability of certain sensitive decisions. They exist to produce good and right answers, but the criteria against which they work remain an enigma. An ethics committee can use academic bioethics by quoting a suitable study to justify the solution the members have already come up with. Academic bioethics has produced, and continues to produce, an endless flow of incommensurable moral views, and the discipline cannot, as a whole, provide a unified starting point for administrative or political decision-making. To choose among the different views and approaches presupposes an ideological choice.

To give legislative or advisory powers to a group of people who have been selected politically, randomly or by acquaintance in issues where there is no general agreement and where decisions have the potential of affecting the lives of many people, seems to me an issue that merits further public discussion. Unfortunately, academic bioethical endeavours cannot, at least at the moment, provide us with a credible alternative.

The Limitations of Academic Knowledge

At least the following claims can be made against philosophical attempts to solve bioethical problems. First, it can be said that ethics exceeds rational arguments and that, therefore, no matter how academically waterproof an argument to support a particular solution might be, it can still be wrong. Secondly, it is impossible to say which of the many normative philosophical frameworks should triumph. A consequentialist argument can always be challenged by stating that there are some absolute limits to what we are allowed to do. These absolute limits, in their turn, can be contested by arguing that there are no such things as absolute limits, and that even if there were, observing them could produce grave suffering. And, of course, then there are the questions of setting those limits: where exactly should they be, and who gets to define their place? Disagreement concerning the most profound premises makes all normative philosophical approaches to bioethics vulnerable to criticism. This is perhaps even more prominent in the case of the religious and theological attempts to solve bioethical questions. If the faith presupposed is not shared, then all subsequent arguments become meaningless.

The relevance of sociological and other social science studies in real-life bioethics can also be questioned. Even if we thought that they could inform us about people's feelings and attitudes, these feelings and attitudes can be ignored on several grounds. It can be argued that the issues behind bioethical problems are so complicated that ordinary people cannot be trusted to understand them, and that, therefore, their ethical views on the matter need not be taken into account. It can also be said that the general public has been wrong before, or that ethics is simply not a matter of general opinion in the first place. A further problem is that qualitative studies that have become popular in recent years are not even meant to be generalisable, and that sociology in general does not claim to produce facts, but instead theoretically enlightened interpretations [6].

Lawyers working in the field of bioethics can tell us about existing legislation, both national and international, and about how new medical and biotechnological breakthroughs will be dealt with within the current legal systems. They can also predict new laws, and the ramifications these might have. In sum, they provide useful information for political and bureaucratic purposes, but this is not usually thought to have anything to do with ethics as such, at least not in the normative sense. It is only those who think that laws are always good as they are who could hold such a view. Others should recognise the possibility of bad, harmful and unjust laws.

Moral philosophers, sociologists and lawyers with relatively similar normative views can work together to provide well-founded ethical opinions regarding important issues. The relevance of these statements can, however, always be questioned. Doubt can be cast on the results by other multidisciplinary groups with different moral convictions, and by those who hold that ethics is not a matter of well-formulated arguments, but rather something that all good and right-minded people intuitively share. Those who hold the latter view usually employ an abstract 'we' to refer to all those people who allegedly have the sensitivity to know the truth about ethics. To those who do not share the convictions, this linguistic attempt to convince others can come across as rather pompous.

Grounds for Dismissing Academic Studies

The significance of the different disciplines involved in bioethics can be questioned both if they pursue their own ends, and if they work towards a common goal. From the viewpoint of an ethics committee trying to overcome a difficult bioethical problem, these disciplines can be useful, but their role can never be exhaustive, or conclusive.

A philosopher cannot really say more than '*if* we believe this, *then*...' Although this normative limitation is explicit only in what I have called the philosophical method of doing bioethics, it is implicit in all the other approaches as well. Philosophers usually acknowledge that their normative work is built on a few chosen premises that cannot be further justified. Their analysis may be accepted if all the premises are shared, and if it is believed that ethics has something to do with rational arguments and coherence. Unluckily for philosophers, many people seem to be quite content to hold incoherent views on ethical issues, and it is very difficult to argue that their approach is somehow decisively inferior to the insistence on coherence. If ethics is believed to involve something more profound than rational arguments, this belief cannot really be shaken by rational arguments.

A sociological study can show a committee that if social scientists use this or that theoretical framework, they can reach results that seem to suggest that some people hold some views on the matter at hand. It is, however, unclear what should be made of these findings. If they are not in accordance with the views of the committee, its members can question the methods of the study and the relevance of its results to any further application. It can also be argued that people's opinions do not really matter – at least when they are thought to be wrong – because they do not adequately understand the issues involved.

It seems clear to me that academic bioethics is a mere means to institutionalised bioethics. Those studies that are consistent with the opinion of those in power, whether sociological, legal, anthropological or historical, can be used to strengthen the political case, and those that are opposed can simply be dismissed. Philosophical arguments for the right cause will be hailed as true, while the 'wrong ones' will be ignored or rejected as failing to encapsulate the essence of true morality, or as having erroneous premises. Sociological results can be useful in a number of ways. They can indicate the limits of people's tolerance, knowledge of which is vital to any political entity's survival. Further, they can suggest the soft spots in people's views, knowledge of which becomes handy, for instance, when trying to implement a new policy that is likely to face opposition. Also there are virtually no limits as to how to interpret and apply the findings of a qualitative study. For practical purposes, it is the lawyers who are the most important source of information for institutionalised bioethics. Ethics committees and other equivalent bodies can, and frequently do, go against public opinion and sound philosophical arguments, but the limits set by law still bind them. Lawyers are of use in knowing the existing laws and their implications, as well as in being capable of predicting future developments in law.

The Regulative Committees

What then could be said about the justifiability of various ethics committees as typical examples of institutionalised bioethics? There are roughly two kinds of committees, those that deal with everyday ethics issues, such as clinical ethics committees in hospitals; and those that give recommendations to governmental bodies, and might even have a legislative role of their own. The clinical-ethics type of committees are mainly interested in legality and good professional conduct, and can be seen as guardians of existing rules. I am, however, more interested in those who make the rules, and I will focus in the latter part of this article on them – what their role is, what it should be, whether this role be justified, and if it can, how [7].

As I have tried to show, there is no consensus on what is moral. Academic bioethics produces different, and incommensurable, theories, norms and proposals. This is also the case outside academia: people have different views on what is ethical and what is not. This is why we have bioethical problems. If everyone agrees on the wrongness of something, it is not thought of as a moral problem, it is perceived as a moral wrong. Unfortunately, most of the issues arising from medicine and the biosciences are such that people's ethical opinions about them differ. Who, then, should be trusted with the regulation of bioethical issues?

Finnish National Ethics Committees – an Example of the Problems

I would think that in liberal democracies bioethical issues, like all issues that have a bearing on many people's lives, should be settled through democratic processes. This is not, however, what happens in ethics committees and advisory boards. As I have recently studied the situation in Finland in some detail, let me cite it as an example [8]. There is no clear procedure as to how the members of the national ethics committees are selected. The unofficial and off-the-record answer given seems to be 'by acquaintance'. There are some legal regulations regarding which professions should be represented in the committees, but the ethical views of those eligible are not made an issue. Various public institutions, including universities, can nominate their representatives, but no written protocol regulates the weighing of the nominations. In practice it seems that those already in the committees can pick the new members on any grounds they choose. Eventually all the appointments are officially approved by the Council of State, but this is a mere formality.

There are also surprisingly few rules as to how the ethics committees are expected to reach their ethical opinions. This leads to an obvious lack of transparency both in the selection of the committee members and in the actual decision-making process. Apart from the representation of various expert groups, required by law, the only explicit criterion for committee membership is independence, or the state or condition of not being under undue influence. Even this minimal requirement is, however, far less clear than it might seem.

The simple reading arguably is that committee members should not stand to benefit financially or in other obvious ways from the ethical evaluations given. That is, for instance, that a member of the ethics committee, who also happens to be an employee of a pharmaceutical company with a new research plan, should not sit in the room when the ethical acceptability of that research proposal is discussed. But other than the mundane issue of financial benefit, what could 'independence' and 'not under undue influence mean'? In ethical issues we have strong opinions for and against. These might be the result of long and careful self-reflection, or they could be something that we have unthinkingly carried with us since granny's childhood teachings, or views that we have picked up at school, in church or during summer camp. The strongest of our ethical opinions are usually those we share with people close to us. What can independence here mean? Can we be unduly influenced by the religious sect we belong to? How about being unduly influenced by one's granny? The problem is that in ethical issues we are never impartial – that we are always influenced by someone or something. It can be argued that the convictions of those sitting on ethics committees form a far more accurate indicator of the type of ethics practised by them than, say, their fields of expertise.

Conclusion

I would like to conclude this short article with a reminder of the state and nature of bioethics today. Many academic bioethicists believe that they have found the good and the truth, but luckily they do not, as a rule, have the power to implement their views as official policies. To the credit of academic bioethics, it should be observed that the views presented in this forum are usually subject to peer-review and critical scrutiny. When it comes to ethics committees which act as official advisory boards for governmental bodies, and sometimes even have legislative power themselves, the situation is, however, strikingly different. There is very little room for external peer-review, but often all the power needed to implement the group's views as official policies. With no academic or general consensus on morality and ethics, and with committees that are often non-democratically selected, the regulation of right and wrong seems to be on shaky grounds. I would like to know what the *ethics* that ethics committees are supposed to promote is, and how do we know that the right people have been appointed to do the job? I cannot think of any other answer than that these issues should be open to public discussion before even more power is entrusted in the hands of ethics committees. I would also like to see academic bioethicists, whether descriptive or normative, trying to make it more difficult for their expertise to be randomly dismissed in institutionalised bioethics.

References

[1] For the different methodological approaches to bioethics see e.g. Bennett, R, and Cribb, A, (2003) 'The Relevance of Empirical Research to Bioethics: Reviewing the Debate' in Häyry, M, and Takala, T, (Eds), *Scratching the Surface of Bioethics*, New York and Amsterdam, Rodopi , 9-18
[2] Beauchamp, T, and Childress, J, *Principles of Biomedical Ethics* (orig. 1979) [5th ed.]. Oxford, Oxford University Press, 2001; Gillon, R. (Ed.), (1994) *Principles of Health Care Ethics*, Chichester: John Wiley & Sons
[3] For analysis of this approach see e.g. Kaplan, A, (1983) 'Can Applied Ethics Be Effective in Health Care and Should it Strive to Be?' *Ethics* 93 pp.311-319
[4] See e.g. Jonsen, A, and Toulmin, S, (1988) *The Abuse of Casuistry*, Berkeley, University of California Press
[5] For a description, see Häyry, M, (1994), *Liberal Utilitarianism and Applied Ethics*. London and New York, Routledge
[6] See e.g. Levitt, M, (2003) 'Better Together? Sociological and Philosophical Perspectives on Bioethics', in Häyry, M, and Takala, T, (Eds), *Scratching the Surface of Bioethics*, New York and Amsterdam, Rodopi, pp. 19-27
[7] The distinction is not always self-evident or clear-cut, but is still used here to separate these different kinds of committees when their role is clearly different. For the grey area in between see e.g. Wilson, R F, (2002) 'Rethinking the Shield of Immunity: Should Ethics Committees Be Accountable for their Mistakes?' *Healthcare Ethics Committee Forum* 14 pp.172-191
[8] Takala, T, and Häyry, M, (2002) 'Ethics committees in Finland: their levels, methods, and point'. *Notizie di Politeia* XVIII – N. 67 pp.60-64

Empirical Investigations of Moral Development

Michael NORUP
Department of Medical Philosophy and Clinical Theory
Panum Institute, Copenhagen

The concept of moral development is tightly connected to the name of Lawrence Kohlberg who for the last three decades has been the leading exponent of the cognitive-developmental approach to the psychological study of morality. This approach has its roots in Piaget's structural approach to child development [1] and more particularly in his study of the moral judgement of the child [2]. But Kohlberg's impact goes beyond the narrow area of child development. His follower James Rest has devised a test (The Defining Issues Test or DIT) to assess moral development. This test, which can be group-administered and computer scored, has given rise to a whole field of research studying morality and factors influencing moral development. According to Rest, the number of studies using the DIT totals well over 1,000 and it has been used in more than 40 countries [3]. A good part of these studies has aimed at investigating the moral development of students and of different professional groups e.g. nurses, physicians, teachers, journalist etc.

Rest's theoretical views are to some extent different from Kohlberg's but their general outlook is the same. There are other more dissenting positions originating from this tradition, such as that represented by Carol Gilligan, who claims that there are significant gender differences in the approach to morality [4]. But Kohlberg's theory is still the point of reference for a whole research tradition.

This in itself is sufficient reason to make it interesting for anyone concerned with the psychological study of morality. The fact that it is based on a series of questionable metaethical assumptions and that it draws some rather controversial normative conclusions on the basis of its results only adds to this interest. A discussion of these issues can help to clarify the relation between moral psychology and normative ethics.

In this paper I will give a short presentation of Kohlberg's theory and of some empirical findings related to the theory. After that I will present Kohlberg's attempt to establish a normative argument on the basis of his research. Finally I will discuss the relevance of philosophical consideration to the empirical study of morality.

Kohlberg's Theory of Moral Development

Kohlberg takes the problem of the justification of moral education as the starting point in his exposition of the philosophy behind his theory [5]. He claims that there are two traditionally accepted views of this problem, but that they both mistakenly accept an untenable doctrine of value relativism. This means that neither of them can offer a justification for their educational ends. In contrast to this, Kohlberg presents his own view as a solution to the relativity problem, which can generate substantial educational objectives, and which is both philosophically and psychologically justified.

In the following I will give a brief sketch of the two traditional views as they are presented by Kohlberg. This sketch is not intended to give a fair description of these views, but only to clarify what sort of problems Kohlberg intended his theory to solve.

The two mistaken views are called respectively *the Cultural Transmission Ideology* and *the Romantic Ideology* [6].

The Cultural Transmission Ideology is said to have its roots in the classical academic tradition of western education, but in more modern variants it is also connected to forms of educational technology inspired by behaviouristic psychology. According to this view the end of education is internalisation of the values and knowledge of the culture. The underlying value premise is one of *social relativity*. The values characteristic of the culture or nation are assumed to be arbitrary. But they make up a consistent set, and as there is at least some consensus about them, adjustment to the culture or achievement within it might as well be taken as the educational end.

The Romantic Ideology seems to be the polar opposite of this. It rests on thoughts connected with Rousseau and is currently represented by followers of Freud and Gesell. According to this view the most important aspects of development come from within. The pedagogical environment should be permissive enough to allow the inner "good" to unfold and the inner "bad" to come under control. But "good" and "bad" are not conceived in ethical terms. Rather they are derived from matters of psychological fact about mental health and happiness. The aim of education is to stimulate individual growth and this is best achieved by giving children freedom to be themselves. There are no common values and the teacher is not entitled to impose his own private outlook on others.

Although these two ideologies end up at opposite extremes, they are both, according to Kohlberg, grounded on value relativism and are therefore unable to justify their educational aims and are bound to end in confusion. *The Cultural Transmission Ideology* leads to "indoctrination, which is incompatible with the conception of civil liberties that are central not only to American democracy, but to any just social system" [7]. *The Romantic Ideology* leads to a refusal to impose the values that the teacher holds to be important on the child, even though these values are the foundation for the teaching.

The *cognitive developmental* or *progressive* approach avoids this relativity problem by proposing a theory of moral development, where the psychologically more mature is also morally better and where the educational goal is the eventual attainment of a higher level or stage.

It is, however, not entirely clear what Kohlberg's argument against relativism is. Following Brandt [8], he distinguishes between ethical relativism and cultural relativism, where cultural relativism is the belief that moral principles are culturally variable in a fundamental way, while ethical relativism is the belief that such divergence is logically unavoidable – that is, that there are no rational principles and methods that could reconcile observed divergences in moral beliefs.

The first of these claims seems to me to be strictly empirical. The other seems to be metaethical. According to Brandt they are distinct in the sense that one who accepts the first can deny the second, while one who accepts the second must also accept the first. It is possible to be a cultural relativist without being an ethical relativist, but it is not possible to be an ethical relativist without being a cultural relativist.

I think Brandt is right in this contention. If people factually share the same principles, and do so for rational reasons, it is hard to deny that it is possible to reach rational agreement. But it is crucial that they have rational reasons for their common beliefs and whether reasons are rational is a philosophical, not an empirical question. If it merely happened to be the case that everybody shared the same set of principles as a matter of psychological or biological fact, this would neither refute ethical relativism nor support the ethical principles people held.

This, however, does not seem to be Kohlberg's view. Rather, he seems to accept Brandt's statement about what it takes to be an ethical relativist, without the rationality proviso, and thinks that if we could disprove cultural relativism we could also disprove ethical relativism and support a certain normative view. He claims that most other social scientists have been confused and biased in favour of both ethical and cultural relativity, thereby being led to argue that science cannot find moral development or evolution. In contrast to this he, by taking ethical non-relativism as his starting point, has been able to find universal categories of moral judgement and basic moral principles that prescribe universal human obligations.

On the basis of his empirical studies he has defined a universal sequence of 6 moral stages characterised not by their *content* but by their form, i.e. by the way morality is conceived. These stages may be considered as separate moral philosophies. They are distinct views of the socio-moral world.

The 6 stages are organised on three levels. Level I is called "Preconventional morality". It includes stage 1 and 2. "Good" and "bad" are interpreted in terms of physical or hedonistic consequences as punishment and reward. Stage 1 is distinguished by an obedience and punishment orientation; stage 2 by a naïvely egoistic orientation. Level II is called "Conventional morality". It includes stage 3 and 4. In stage 3 there is morality of maintaining good relations and seeking the approval of others; in stage 4 an authority and social-order-maintenance orientation is dominant. Level III is termed "Postconventional" or "Principled morality". It includes stage 5 and 6. Here there is a clear effort toward autonomous moral principles. Stage 5 is a morality of contract, of individual rights, and of democratically accepted laws. Stage 6 is a morality of individual principles of conscience. (For details see table 1, in appendix.)

The stages are claimed to live up to the following four criteria defined by Piaget [9]:
1. Stages imply a distinction or qualitative difference in structure (modes of thinking) that still serves the same basic function (for example intelligence) at various points in development.
2. These different structures form an invariant sequence, order or succession in individual development. While cultural factors may speed up, slow down, or stop development they do not change its sequence.
3. Each of these different and sequential modes of thought forms a "structured whole". A given stage response on a task does not just represent a specific response determined by knowledge and familiarity with the task or tasks similar to it; rather, it represents an underlying thought-organisation. The implication is that various aspects of stage structures should appear as a consistent cluster of responses in development.
4. Stages are hierarchical integrations. Stages form an order of increasingly differentiated and integrated *structures* to fulfil a common function. Accordingly, higher stages displace (or rather integrate) the structure found at lower stages.

Kohlberg assessed people to be at different stages through his Moral Judgement Interview. This is a semi-structured interview in which he presented subjects with a series of hypothetical moral dilemmas and asked them about them, particularly attending to their rationale for saying why one line of action was more morally justified than another. An example of such a dilemma is the following, which is called the Heinz dilemma:

In Europe, a woman was near death from a very bad disease, a special kind of cancer. There was one drug that the doctors thought might save her. It was a form of radium that a druggist in the same town had recently discovered. The drug was expensive to make, but the druggist was charging ten times what the drug cost him to make. He paid $200 for the radium and charged $2,000 for a small dose of the drug. The sick woman's husband, Heinz, went to everyone he knew to borrow the money, but he could get together only $1,000, which was half of what it cost. He told the druggist that his wife was dying and asked him to sell it cheaper or let him pay later. But the druggist said, "No, I discovered the drug and I'm going to make money from it". Heinz got desperate and broke into the man's store to steal the drug for his wife.

Should Heinz have done that?

Possible answers to this question could be [10]:
a. It really depends on how much Heinz likes his wife and how much risk there is in taking the drug. If he can get the drug in no other way and he really likes his wife he'll have to steal it.
b. I think that a husband would care so much for his wife that he couldn't just sit around and let her die. He wouldn't be stealing for his own profit, he'd be doing it to help someone he loves.
c. Regardless of his personal feelings, Heinz has to realise that the druggist is protected by the law. Since no one is above the law, Heinz shouldn't steal it. If we allowed Heinz to steal, then all society would be in danger of anarchy.

Such responses could be compared to examples and criteria in a scoring guide. In this case the 3 responses would probably be scored as examples of stage 2, 3 and 4 reasoning respectively. During a full interview typically about 50 matches between the subjects responses and the manual's examples will be found and using summary rules it will be possible to assign the subject an overall stage.

Rest's Defining Issues Test (DIT) [11] is different from this. It has the form of a multiple-choice test. The subjects are presented with 6 dilemmas, some of which are the same as in Kohlberg's moral interview, (e.g. the Heinz dilemma), and some which are different. The subject's task is, however, not to produce reasons for a particular line of action, but to evaluate among 12 items given for each dilemma the extent to which they raise important considerations for deciding the case. The subject is asked to rate the relative importance of each item on a 5 point scale, from great importance to no importance, and to rank which of the 12 items is the most important, the second most important, and so on. The items are written as fragments of a way of thinking about a dilemma typical for each stage. The most frequently used score is the so called P-score (Principled Score) which is based on the relative importance subjects give to items representing stage 5 and 6: principled moral thinking. Rest claims that there is a correlation in the 0.6 to 0.7 ranges between results obtained by the DIT and results obtained with Kohlberg's interview method [12].

What Are the Empirical Findings?

It is very hard to assess the empirical evidence of the theory as both the description of stages and the scoring system has changed through the years [13]. However one attempt has been made by Bergling [14]. He has isolated some basic claims of the theory and some hypotheses derived from the postulates, which he has then tested on the basis of empirical results. The hypothesis that stage acquisition is sequential and progressive was tested on results from 5 longitudinal studies. It was supported for stage 1-4 in children, but rejected for adults. The hypothesis that the acquisition of stages occurs stepwise, that is, that no stages are skipped, was supported for children in stages 1-3 but rejected for adolescents and adults.

The hypothesis about cultural universality of the scheme of moral development was not supported in any strong sense. In studies from Bahamas and British Honduras no representatives of stage 4-6 were found and in a study from Nigeria Stage 3 was found to be underrepresented. Findings from western industrialized countries support stage 1-4.

In contrast to this, Rest reports that studies with the DIT in 40 countries showed that score increased, not with age, but with degree of formal education [15]. This, however, can only be taken as very weak evidence of cultural universality. Only subjects with a certain level of reading skills can take the DIT. It is likely that acquisition of reading skills and formal education in itself has some homogenising effect on the way people think about morality and on the sorts of judgement they find attractive.

The claim that each stage forms a "structured whole" has been assessed by Rest [16]. He points out that there are at least two problems involved in assigning subjects to stages: firstly, the gradual nature of acquisition of stages, and secondly, the instability of responses during transitional phases. Further he points out that the usual procedure for classifying subjects in accordance with stage type is to look for the subject's quantitative distribution of responses and to designate the type with most responses as the subject's own stage. The number of scores at a particular stage seldom exceeds 50%. This means that even though the stage theory is qualitative by nature the actual scoring method is quantitative. These problems have led Rest to dismiss what he calls a simple stage model and to propose a new, more complex developmental model [17]. On this model we should not ask: "What stage is a person in?" but rather, " To what extent and under what conditions does a person manifest the various types of organisations of thinking?" This model is probabilistic. It describes an individual's stage use in terms of a probability of each stage for a specific task.

The difference between Rest's and Kohlberg's stage models seems to be profound and it would be interesting if Kohlberg's idea of discrete, qualitatively different stages could obtain empirical support, but I will not go further into this question in the present context.

The Claim to Moral Adequacy of Higher Stages and the Problem with Stage 6

It is central to Kohlberg's attempt to justify moral education in terms of moral development that higher stages can be seen as better than lower ones in a morally relevant sense.

In the article "From Is to Ought: How to Commit the Naturalistic Fallacy and Get Away With It in the Study of Moral Development" [18], Kohlberg defended this view in a rather bold form. But in the more recent book "Moral Stages: A Current Formulation and a Response to Critics" [19], most of the more controversial claims of this article were retracted. One of the main reasons for this was that the status of stage 6 was changed.

The empirical evidence for stage 6 has always been weak, and as it was not found in any subjects in longitudinal studies in the United States, Turkey or Israel, it was abandoned as an empirically identifiable form of moral reasoning. This means that it is not included in the latest scoring system. But that doesn't mean that it is no longer part of the theory. It has obtained the status of a theoretical construct, a rational endpoint which is thought to be necessary for the theory and which Kohlberg was working to define until his death in 1987. The final attempt to propose a formulation was present in article named " The Return of Stage 6: Its Principle and Moral Point of View" [20].

In order to discuss the modifications of Kohlberg's theory that result from the disappearance of stage 6 as an empirical stage, it is necessary to look at his former claims about stage 6 and his claim that higher stages are more morally adequate than lower ones.

These claims rest on a thesis of *isomorphism* or *parallelism* between psychological theory and normative ethical theory. What is meant by this is roughly that the reasons that explain why people move upward from one stage to the next and the reasons why higher stages are more morally adequate are somehow the same. The thesis is not easy to make precise on the basis of Kohlberg's writings. I will not try to explain it in detail in this context, but merely outline its main structure.

One part of the argument is the claim that higher stages are psychologically more adequate than lower. The movement from one stage to the next is said to be induced by cognitive conflict and this is stimulated by opportunities for role-taking. This process leads to stages, which are successively more integrated and differentiated, and which give the subject better conflict-resolving abilities. Higher stages are in this way better conceptual tools for making sense of the world and deriving guides for decision-making.

At the same time higher stages are more complex and cognitively difficult. This is shown by empirical studies [21]. Adolescents asked to reinterpret statements higher than their own level distorted them into ideas at their own level or below, but they had no difficulty in comprehending arguments at or below their own stage.

But this is not all. Besides being an order of increased cognitive difficulty and inclusiveness, where people on higher stages are able to solve more problems than people on lower, Rest also found that stages form an order of perceived moral adequacy. When adolescents were asked to rate moral statements according to their adequacy or persuasiveness, they gave the highest rating to statements at their own stage or above, to the extent that they could comprehend them. This means that people prefer higher stage thinking to lower, as far as they can comprehend it. This preference, it seems, is part of the explanation of their change. Further, according to Rest [22], the reasons the subjects give for preferring higher stages are the same reasons Kohlberg gives in his theory for why higher stages are better.

To Rest this seems to be enough to establish that higher stages are indeed morally better. Kohlberg however seems to be aware that such preferences are not sufficient to establish the moral adequacy of higher stages. Even though he states that: "The scientific theory as to why people factually *do* move upward from stage to stage, and why they factually *do* prefer a higher stage to a lower, is broadly the same as a moral theory as to why people *should* prefer a higher stage to a lower " [23], he also gives some other arguments for his thesis of parallelism. The arguments are based on structure and formal considerations. [24]. It is part of the structural theory that the formal properties of *integration* and *differentiation* are criteria for psychological adequacy. As these criteria are also applicable to moral reasoning it is possible to say that higher stages are more structurally adequate, not only psychologically, but also morally. In addition to this the formal criteria of adequacy – levels of *differentiation* and *integration* – are said to be parallel to formalistic moral philosophy's criteria of *prescriptivity* and *universality*. Further, it is stated that "These two criteria combined represent a formalistic definition of the moral, with each stage representing a successive differentiation of the moral form from the nonmoral and a more full realization of the moral form" [25].

As judgements become increasingly integrated, they become increasingly universalisable. This may be clarified by the following: moral judgements concerning the value of human life are said to become increasingly prescriptive as one move up through the stages in that "the moral imperative to value life becomes increasingly independent of the factual properties of the life in question. First, people's furniture becomes irrelevant to their value; next, whether they have loving families; and so on (it is correspondingly a series of differentiation of moral consideration from other considerations). In parallel to this there is a "movement toward increased universality of moral valuing of human life. At stage 1, only important people's lives are valued; at stage 3, only family members; at stage 6, all life is to be morally valued equally" [26].

In summary, Kohlberg's claim to the moral adequacy of higher stages of moral development rests on two claims.
1. The claim that the explanation of why people factually do move upward through stages and the explanation why they should is roughly the same, and
2. The claim that there is a formal parallelism between psychological and moral philosophical criteria of adequacy.

None of these claims are convincing, but this is not an issue I will pursue here. Good arguments are presented by Siegel [27]. Instead, I will go on with my presentation of Kohlberg's view.

According to the theory, stage 6 is the ultimate endpoint of moral development. It is maximally integrated and differentiated, reversible and equilibrated. Its principles are not only designed to be acceptable to all rational people, but all those who use stage 6 reasoning will eventually agree on the "right" solution in concrete situations. The stage represents the highest level in Kohlberg's developmental theory, and it defines the lower stages as steps in which the characteristics of stage 6 are only present in a rudimentary or incomplete form. Stage 6 reasoning is based on a notion of justice as reversibility as conceived by Rawls [28]. But this doesn't mean that the validity of stage 6 principles depend on the validity of Rawl's theory or on any other theory justifying these principles. To Kohlberg the relation is rather the reverse. The adequacy of a moral theory derives from the principles it supports. The central achievement of Rawl's theory is that it represents the first clear and systematic justification of stage 6 principles and methods.

As stage 6 was abandoned as an empirical stage, Kohlberg could no longer claim to have proved that there was a highest and most morally adequate empirically verifiable stage of moral reasoning, and he presented a new revised version of his theory and admits in response to Gilligan [29] and others that there are aspects to morality other than justice. This, together with influence from Habermas, led him to redescribe his theory as a "*rational reconstruction of ontogenesis of justice reasoning*".

One of the major changes in the new version is that the thesis of *parallelism* is substituted with a thesis of *complementarity*. On this thesis a statement as to the normative adequacy of an ethical theory cannot be tested by or be shown to imply, empirical truth claims. However, "the success of an empirical theory which can only be true or false can function as a check for the normative validity of a hypothetical reconstruction of moral intuitions" [30].

But even though this thesis is less blurred, (to use Kohlberg's own word), than the thesis of parallelism, it is not entirely clear. What exactly is meant by "function as a check" is not made explicit. If it is interpreted in a weak sense, as a claim that lack of empirical support casts some doubt on our normative theories, it may be credible. If people showed no tendency to develop in the direction of our theory, while they at the same time showed evidence of moral thinking totally at odds with it, this might give us a reason to rethink our normative position.

If, on the other hand, the thesis is interpreted in a strong sense, if it is thought that empirical data can somehow falsify a normative theory, it is highly implausible. It is a part of many moral theories, that they make claims about what we ought to do that we are not psychologically disposed to follow. This does not show that these theories are wrong but only that there are problems connected with their application. At the same time this can hardly be Kohlberg's view. As Boyd points out [31], Kohlberg shows little evidence of having given up his faith in the normative validity of stage 6 in spite of its lack of empirical support. Rather he seems to be inclined to discuss how our educational efforts may be altered to effect a change, so that people would in fact evidence the approved normative position.

Another change is that it is no longer claimed that there is a single principle used at the current highest empirical stage, nor that this principle is the principle of justice or respect for persons. Instead stage 6 is seen as a rational ideal of moral development. The normative claims for the moral adequacy of higher stages are thought not to be necessary in any strong sense, for the psychological study of moral stages per se. Their status is rather hypothetical, as part of a psychological-normative theory, which is a "rational reconstruction of development". The claim that higher stages are morally more adequate than lower ones must be made if stage development is to be taken as an aim of education, but the moral stage framework can be a fruitful tool for scientific work without this claim [32].

On the other hand it is stated that there are certain metaethical assumptions that it is necessary to endorse to begin a psychological study of morality using the stage concept. These

assumptions are:
(i) Value relevance of definitions of the moral
(ii) Phenomenological definitions of morality (as opposed to behaviourism)
(iii) Moral universality (as opposed to cultural and ethical relativism)
(iv) Prescriptivism (as opposed to descriptivism or naturalism as interpretations of moral judgements)
(v) Cognitive (as opposed to emotivism)
(vi) Formalism as defining the nature and competence of moral judgements (as opposed to definitions in terms of content)
(vii) Principledness as the rule or governance of moral judgement (as opposed to act theory)
(viii) Constructivism (as opposed to either empiricism or apriorism) 'To these metaethical starting assumptions is added a more normative or substantive assumption, namely that justice is primary, that moral problems as dilemmas are fundamentally problems of justice.' [33].

I will not give any further description of these assumptions. They are only mentioned because the idea of making such assumptions will be discussed later.

The Relation between Moral Psychology and Ethics

Turning to the problem about the relation between moral psychology and ethics, we can ask two questions:
1) What can the study of moral psychology do for normative ethics?
2) What can philosophical ethics do for the psychological study of morality?

The answer to the first of these questions can, I think, be quite short. Neither Kohlberg's thesis of parallelism nor can his thesis of complementarity establish the claims to the moral adequacy of higher stages. Moral psychology can neither solve the problem of justification of moral education nor the problems about the plausibility of normative ethical theories. We can only decide educational ends on the basis of what we think is valuable in the context, and then, if necessary, turn to appropriate research to find out how the attainment of these ends can best be facilitated.

But while empirical studies of moral psychology cannot solve philosophical questions, they can be of interest in at least two other ways. The first is general. Among the many various sorts of empirical knowledge relevant to normative ethics, knowledge of people's thinking about morality is likely to play a prominent role. The second is connected to the teaching of ethics and to the application of normative principles. If we wish to teach people or make them understand or accept certain normative positions, it is of value to know how they conceive of normative problems.

Measuring methods like those devised by Kohlberg and Rest might also play a role in this context. But if we want to use them as measures for educational success we must show, by way of philosophical arguments, that there are good reasons to think that each stage in the developmental sequence is better than its predecessor. Such an argument cannot restrict itself to the claim that higher stages are actually preferred, that they are more integrated and differentiated or that they are more impartial, just and reversible. It would also have to show the respects in which attainment of higher stages enables people to make better moral judgements or to what extent it provides them with the sort of characteristics we appreciate. It is far from clear that this is possible for Kohlberg's or Rest's stages. As Locke points out [34], some of us might prefer a world of stage 3 reasoners motivated by interpersonal concerns of caring to a world of stages 4 advocates of law and order or to a world of stage 5 social utilitarians.

The second question can be reformulated as a question about how the psychological study of morality should proceed and which philosophical assumptions are necessary. A full answer to this question will far exceed the scope of this paper and I will therefore only outline some proposals and state my position.

The first thing I want to point out is that if we want to use Kohlberg's concept of moral stages, we should give up the notion of a most adequate structural endpoint and the claim that such an endpoint is needed to define the developmental sequence of moral judgement. This means that we should give up stage 6. Kohlberg maintained stage 6 as a theoretical construct, because this stage was seen as necessary to define the nature and endpoint of the kind of development under study. He conceived his theory as a rational reconstruction of the ontogenesis of justice reasoning, that is, as a theory describing the developmental logic inherent in the development of justice reasoning with the aid of the normative criterion of stage 6, which is held to be the most adequate stage of justice.

There are several problems with this position. As Puka points out [35], by maintaining stage 6 as the optimal endpoint of moral development, Kohlberg needlessly rendered himself partisan in the philosophical debate between deontologists and teleologists. Further, his description of stage 6 reasoning does not establish that this way of thinking gives the sort of solutions to moral problems that he postulates. It does not seem to be true that people on this stage will always be able to reach agreement or find just solutions to moral problems. And even if they could, this does not tell us that natural stages of moral development can be described satisfactorily in stage 6 terms. Instead of maintaining a dubious philosophical conception of morality as an endpoint in moral development, it would be more interesting to search for data regarding the possible existence of other natural stages beyond stage 5. This could be done by hypothesising a variety of possible structures for interpreting moral judgement that seems to transcend stage 5.

Besides this, I think that we should reconsider the metaethical assumptions that Kohlberg thinks psychologists using the stage concept must necessarily endorse. The assumption of phenomenalism should, in my opinion, be maintained as central. It defines a special approach to moral psychology, which I consider to be the most interesting. But having made this assumption, I think, following Blasi [36], that the others, to the extent that they are considerations derived from specific philosophical theories, should not be used as starting assumptions for empirical research. They tend to restrict the field of inquiry unnecessarily and to tempt the researcher to look for what *should* be rather than for what *is*. A better use of such philosophical considerations would be to take them as possible guidelines for the construction of hypotheses and to try to formulate them in empirically testable terms. At the same time I don't think moral psychology should restrict itself to study morality as it is understood by philosophers. There may be areas frequently considered to be morally relevant in ordinary language and common understanding which are at odds with philosophical understanding, but for a psychology interested in people's actual thinking, such areas cannot be ignored.

One of the most problematic of Kohlberg's initial assumptions is the assumption of value relevance of definitions of the moral. This seems to imply that the researcher must take a stand on morality that can be philosophically justified, and that he must believe that systems of truly moral thought are more or less valid and that movement from one system to another can represent a developmental sequence in distinctively moral thought [37].

This assumption is not only superfluous, but also detrimental. It is not necessary for the researcher to make claims about whether people's moral reasons are more or less philosophically sound. His task is to make the individuals moral thinking understandable on its own terms. If a person makes a stage 3 judgement it should not be evaluated by stage 5 or 6 standards but understood according to the person's own view and understanding of the situation. Otherwise the assumption of phenomenalism is compromised.

The assumption seems to be a reminiscence of Kohlberg's old "is" to "ought" claims. But as far as I can see it can be given up without much loss. If we accept that the aim of developmental theory is to provide *a rational reconstruction of ontogenesis of justice reasoning* our strategy could be to propose different normative ethical theories as conjectures of such reconstructions. These normative theories could then be tested as such conjectures and may be falsified. This could help us to formulate normative theories with the best possible fit to people's actual way of thinking. But it would not, as pointed out earlier, help us to say anything about the philosophical merits of these theories.

Table One
Content of stage

Level and stage	What is right	Reasons for doing right	Social perspective of stage
Level I: Preconventional			
Stage 1 Heteronomous morality	To avoid breaking rules backed up by punishment; obedience for its own sake, and avoiding physical damage to persons and property.	Avoidance of punishment and the superior power of authorities.	Egocentric point of view. Doesn't consider the interests of others or recognize that they differ from the actor's: doesn't relate two points of view. Actions are considered physically rather than in term of the psychological interests of others. Confusion of authorities' perspective with one's own.
Stage 2 Individualism, instrumental purpose and exchange	Following rules only when it is in someone's immediate interests, acting to meet one's own interests and needs and letting others do the same. Right is also what is fair, what's an equal exchange, a deal, an agreement.	To serve one's own needs and interests in a world where you recognize that other people have their interests too.	Concrete individualistic perspective. Aware that everybody has his own interests to pursue and that these conflict, so that right is relative (in a concrete individualistic sense).
Level II: Conventional			
Stage 3 Mutual interpersonal expectations, relationships and interpersonal conformity	Living up to what is expected by people close to you or what people generally expect of your role (son, brother, friend etc). "Being good" is important and means having good motives, showing concern for others. It also means keeping mutual relationships such as trust, loyalty, respect and gratitude.	The need to be a good person in your own eyes and those of others. Your caring for others; Belief in the Golden Rule. Desire to maintain rules and authority which support stereotypical good behaviour.	Perspective of the individual relationship with other individuals. Aware of shared feelings, agreement, and expectations, which take primacy over individual interest. Relate points of view through the concrete Golden Rule, putting yourself in the other guy's shoes. Doesn't yet consider generalized system perspective.
Stage 4 Social system and conscience	Fulfilling duties to which you have agreed. Laws are upheld except in extreme cases where they conflict with other fixed social duties. Right is also contributing to society, the group or institution.	To keep the institution going as a whole, to avoid the breakdown in the system "if everyone did it", or imperative of conscience to meet one's defined obligations (easily confused with stage 3 belief in rules and authority).	Differentiates social point of view from interpersonal agreement motives. Takes the point of view of the system that defines roles and rules. Considers individual relations in terms of place in system.
Level III: Postconventional			
Stage 5 Social contract or utility and individual rights	Being aware that people hold a variety of values and opinion; that most values and rules are relative to your group. These relative rules should mostly be upheld, however, in the interest of impartiality and because there are social contracts. Some nonrelative values like life and liberty, however, must be upheld in any society regardless of majority opinion.	A sense of obligation to law because of one's social contract to make and abide by laws for the welfare of all and for the protection of all people's rights. A feeling of contractual commitment, freely entered into, to family, friendship, trust, and work obligations. Concern that laws and duties be based on rational calculation of overall utility, "the greatest good for the greatest number.	Prior society perspective. Perspective of a rational individual aware of values and rights prior to social attachments and contracts. Integrates perspectives by formal mechanisms of agreement, contract, objective impartiality, and due process. Considers moral and legal points of view; recognizes that they sometimes conflict and find it difficult to integrate them.

Stage 6 Universal ethical principles	Following self-chosen ethical principles. Particular laws or social agreement are usually valid because they rest on such principles. When laws violate these principles one acts in accordance with the principle. Principles are universal principles of justice: the equality of human rights and respect for dignity of human beings as individual persons.	The belief as a rational person in the validity of universal moral principles and a sense of commitment to them.	Perspective of a moral point of view from which social arrangements derive. Perspective is that of any rational individual recognizing the nature of morality or the fact that persons are ends in themselves and must be treated as such.

References

[1] Piaget, J, (1960), 'The General Problem of Psychological Development of the Child', in Tanner, JM, and Inhelder, B, (Eds.), *Discussions on Child Development: A Consideration of the Biological, Psychological, and Cultural Approaches to the Understanding of Human Development and Behaviour*, Vol. 4. New York: International University Press
[2] Piaget, J, (1965), *The Moral Judgement of the Child*, New York: The Free Press (Org 1932)
[3] Rest, J, (1994), 'Background: Theory and research', in Rest, J and Narváez, D, (Eds.), *Moral Development in the Professions*, Hillsdale, New Jersey: Lawrence Erlbaum Associated Publishers
[4] Gilligan, C, (1982*)*, *In a Different Voice*, Cambridge, Mass: Harvard University Press
[5] Kohlberg, L, (1981), *The Philosophy of Moral Development*, San Francisco: Harper and Row
[6] Kohlberg, L, (1981), ibid
[7] Kohlberg, L, (1981), ibid, p.8
[8] Brandt, (1961), cited in Kohlberg, L, (1981), ibid, p.107
[9] Piaget, J, (1960), op cit
[10] From Rest, J, (1994), op cit
[11] Rest, J, (1979), *Development in Judging Moral Issues,* Minneapolis: University of Minnesota Press
[12] Rest, J, (1994), op cit
[13] Colby, A, (1978), 'The Evolution of Moral Developmental Theory', in *Moral Development. New Directions for Child Development,* 2: 89-104
[14] Bergling, K, (1981), *Moral Development: The Validity of Kohlberg's Theory*, Stockholm: Almqvist & Wiksell int.
[15] Rest, J, (1986), *Moral Development: Advances in Research and Theory*, New York: Praeger
[16] Rest, J, (1979), op cit
[17] Rest, J, (1986), op cit
[18] Kohlberg, L, (1981), op cit
[19] Kohlberg, L, (1983), op cit
[20] Kohlberg, L, Boyd, DR, Levine, C, (1990), 'The Return of Stage 6: Its Principles and Moral Point of View', in Wren TE, (Ed.), *The Moral Domain: Essays in the Ongoing Discussion between Philosophy and the Social Sciences*, Cambridge, Mass: MIT Press
[21] Rest, J, Turiel, E, and Kohlberg, L, (1969), Relations between Level of Moral Judgement and Preference and Comprehension of the Moral Judgements of Others, *Journal of Personality* 37: 225-252; Rest, J, (1973), The Hierarchical Nature of Moral Judgement, *Journal of Personality* 1: 86-109
[22] Rest, J, (1994), op cit
[23] Kohlberg, L, (1981), op cit, p.179
[24] In my discussion I mainly follow Siegel, H, (1985), On Using Psychology to Justify Judgement of Moral Adequacy, in Modgil, S and Modgil, C, (Eds.), *Lawrence Kohlberg: Consensus and Controversy,* Philadelphia: The Falmer Press
[25] Kohlberg, L, (1981), op cit, p.171
[26] Kohlberg, L, (1981), op cit, p.135
[27] Siegel, H, (1985), op cit
[28] Kohlberg, L, (1981), op cit
[29] Gilligan, C, (1982*),* op cit
[30] Kohlberg, L, (1983), op cit, p.15
[31] Boyd, DR, (1985), .The Ought of Is: Kohlberg at the Interface between Moral Philosophy and Developmental Psychology', in Modgil, S and Modgil, C, (Eds.), *Lawrence Kohlberg: Consensus and*

Controversy, Philadelphia: The Falmer Press
[32] Kohlberg, L, (1983), op cit, p.65
[33] Kohlberg, L, (1983), op cit, p.66-67
[34] Locke, D, (1985), 'A Psychologist among the Philosopher. Philosophical Aspects of Kohlberg's Theories' in Modgil S and Modgil, C, (Eds.), *Lawrence Kohlberg, Consensus and Controversy,* Philadelphia: The Falmer Press
[35] Puka, B, (1990), 'The Majesty and Mystery of Kohlberg's Stage 6', in Wren TE (Ed.), *The Moral Domain: Essays in the Ongoing Discussion Between Philosophy and the Social Sciences,* Cambridge, Mass.: MIT Press
[36] Blasi , A, (1990), 'How Should Psychologists Define Morality? Or, The Negative Side Effects of Philosophy's Influence on Psychology', in Wren, TE, (Ed.), *The Moral Domain: Essays in the Ongoing Discussion between Philosophy and the Social Sciences,* Cambridge, Mass.: MIT Press
[37] Kohlberg, L, (1983), op cit, p.68-69

Bibliography

Berling, K, (1981), *Moral Development: The Validity of Kohlberg's Theory,* Stockholm: Almqvist & Wiksell int.
Blasi, A, (1990), 'How Should Psychologists Define Morality? Or, The Negative Side Effects of Philosophy's Influence on Psychology', in Wren, TE, (Ed.), *The Moral Domain: Essays in the Ongoing Discussion between Philosophy and the Social Sciences,* Cambridge, Mass.: MIT Press
Boyd, DR, (1985), 'The Ought of Is: Kohlberg at the Interface between Moral Philosophy and Developmental Psychology' in Modgil, S and Modgil, C, (Eds.), *Lawrence Kohlberg: Consensus and Controversy.* Philadelphia: The Falmer Press
Colby, A, (1978),'The Evolution of Moral Developmental Theory' in *Moral Development: New Directions for Child Development,* 2: 89-104
Gilligan, C, (1982), *In a Different Voice,* Cambridge, Mass: Harvard University Press
Kohlberg, L, (1981), *The Philosophy of Moral Development,* San Francisco: Harper and Row
Kohlberg, L, (1985), 'A Current Statement of Some Theoretical Issues', in Modgil, S and Modgil, C, (Eds.), *Lawrence Kohlberg: Consensus and Controversy.* Philadelphia: The Falmer Press
Kohlberg, L, Boyd, DR, Levine, C, (1990), 'The Return of Stage 6: Its Principles and Moral Point of View', in Wren TE, (Ed.), *The Moral domain, Essays in the Ongoing Discussion between Philosophy and the Social Sciences,* Cambridge, Mass.: MIT Press
Kohlberg, L, Levine, FC, Hewer, A, (1982), *Moral Stages: A Current Formulation and a Response to Critics,* Basel: Karger
Locke, D, (1985), 'A Psychologist among the Philosopher. Philosophical Aspects of Kohlberg's Theories', in Modgil, S and Modgil, C, (Eds.), *Lawrence Kohlberg, Consensus and Controversy.* Philadelphia: The Falmer Press
Piaget, J, (1965), *The Moral Judgement of the Child,* New York: The Free Press, (Org. 1932)
Piaget, J, (1960), 'The General Problem of Psychological Development of the Child', in Tanner, JM and Inhelder, B (Eds.), *Discussions on Child Development: A Consideration of the Biological, Psychological, and Cultural Approaches to the Understanding of Human Development and Behaviour,* Vol. 4 New York: International University Press
Puka, B, (1990), 'The Majesty and Mystery of Kohlberg's Stage 6' in Wren, TE, (Ed.), *The Moral Domain: Essays in the Ongoing Discussion Between Philosophy and the Social Sciences,* Cambridge, Mass.: MIT Press
Rest, J, (1973), 'The Hierarchical Nature of Moral Judgement', *Journal of Personality* 1: 86-109
Rest, J, (1979), *Development in Judging Moral Issues,* Minneapolis: University of Minnesota Press
Rest, J, (1986), *Moral Development: Advances in Research and Theory,* New York: Praeger
Rest, J, (1994), 'Background: Theory and Research', in Rest, J and Narváez, D, (Eds.), *Moral Development in the Professions,* Hillsdale, New Jersey: Lawrence Erlbaum
Rest, J, Turiel, E, and Kohlberg, L, (1969), 'Relations between Level of Moral Judgement and Preference and Comprehension of the Moral Judgements of Others' *Journal of Personality* 37: 225-252
Siegel, H, (1985), 'On Using Psychology to Justify Judgement of Moral Adequacy'. in Modgil, S and Modgil, C (Eds.), *Lawrence Kohlberg: Consensus and Controversy,* Philadelphia: The Falmer Press

Theory and Methodology of Empirical-Ethical Research

Lieke VAN DER SCHEER, Ghislaine VAN THIEL, Johannes VAN DELDEN,
Guy WIDDERSHOVEN
Healthcare Ethics and Philosophy
Malden, Netherlands

Practical Turn

Twentieth century practical ethics developed as a result of the conviction that moral philosophy should provide practical guidance in actual moral issues, and of the perception that ethical reflection and theorising was obviously relevant to health care, politics, science, and business. Practical ethics is considered as a turn from abstraction in modern ethical theories [1] to the concreteness of daily practice.

Both principlism and casuistry as forms of practical ethics have developed their own path in an attempt to account for the concreteness of life. These approaches have been elaborated in some detail. While principlism is a form of practical ethics that wants to apply existing theoretical principles to real-life problems, casuistry claims that not only the problems, but the moral rules themselves have their origin in the concreteness of life experience [2]. These approaches have some problems. A problem with principlism is that moral theories are very general and therefore insensitive to the peculiarities which characterize a certain situation in practice. Moreover social research investigating the way in which moral decisions are actually taken, suggests that such decisions are not reached by the application of some ethical method. On the other hand, there is the objection against casuistry that it is pretty arbitrary. It lacks the rules necessary for the interpretation of a particular case, nor is there a procedure to structure moral reasoning. It is not clear, either, whether, and if so, how, casuistic ethics can fulfil the critical and/or prescriptive function of a normative ethics.

Empirical Research in Ethics

Since the last decades of the twentieth century a new step occurred in practical ethics. Ethical works began to appear in which the empirical methods of social scientists were used [3]. There is growing awareness that the study of the lived world and the explicit formulation thereof yields information that is meaningful for ethics, if ethics wants to be relevant for moral practice. The possibilities and impossibilities of such empirical contributions to ethics are being discussed extensively [4]. In the literature we distinguish in the first place the approach in which empirical research and ethics are combined as an interdisciplinary enterprise. Here, the primary function of empirical facts is to inform ethics. This approach focuses upon the relation between descriptive and prescriptive statements [5]. It wants to combine empirical research with ethics, especially the results of life sciences and of social sciences, and maintains the distinction between description and prescription. In concrete terms it comes down to this: science cannot say anything about ends to be achieved, while ethics can never be derived from factual scientific information. While this approach maintains a belief in the critical function of ethics, it does not question the content of normative guidelines.

We distinguish in the second place the ethnographic approach in ethics. Ethnography studies how people actually make ethical decisions and what values they hold. It helps to understand how, and why, people act as they do. This approach is said to make a critical contribution to our understanding of morality. The results of ethnographic research challenge the dogmas and the underlying philosophical model in applied ethics. It is also said to reveal that morality must be understood contextually. This ethical approach however is criticized because in using empirical data, one seems to confuse the descriptive, the analytical-metaethical and the normative domains of ethics. Descriptive and metaethical ethics, so it is argued, cannot provide normative guidelines [6].

Empirical Ethics

In this article we present a relatively new trend in ethical research, called 'empirical ethics'[7]. It is in this type of ethics research that we try to make a step forward in practical ethics, that is to say, in the process of combining ethical theorising with a contribution to handling problems related to practical issues.

At this point in time, there is no clear and commonly accepted definition of empirical ethics. The purpose of this section is to describe empirical ethics by pointing out some characterizations of empirical ethics. Furthermore, we investigate two proposals to connect the empirical with normativity, namely reflective equilibrium and pragmatic hermeneutics.

Theory and practice

A first aspect of empirical ethics is that it aims at developing normative theories which do justice to the complex and - especially - unique character of situations and practices. This means, among other things, that moral reasoning must involve and integrate people's moral intuitions and experiences regarding a given case.

A condition for the possibility of empirical ethics is that theory and practice are mutually integrated. In order to make such integration visible, it may be good to stress the rediscovery of practice in the characterization of empirical ethics which must be seen primarily as a reaction to purely theoretical approaches to ethics. Authors in this field propose that more attention should be given to what people do in practice and what motivates them. Within this perspective, an attempt is made to formulate normative theories by means of an empirical inquiry into the practice to which the theories are relevant [8].

Descriptive and prescriptive aspects

The second point that can be made is that research in empirical ethics is concerned with the *moral* aspects of specific practical problems. In the course of such research, questions are raised concerning the rightness of a certain course of action; what it is to be a good human being or a good thing; what it is that our duty demands that we do, etcetera. We also think that the study of the factual situation in the lived world and its explication yields a philosophy which can be relevant to non-academic practice. The two aspects, both the moral and the factual, are thought to be relevant. But they are difficult to combine if we stick to the epistemological distinction between empirical research into knowable facts and ethics, which is related to values.

From an ethical point of view the distinction may best be characterised in its linguistic form, as the distinction between descriptive and normative statements. It may be useful to point out that, self-evident though it may seem, the distinction between fact and value, between descriptive and prescriptive and, concomitantly, the distinction between is and ought, is indeed the result of a historical process in which science and ethics have increasingly become separated [9]. Our hypothesis is that by abandoning the belief in the fixed gap between descriptive and prescriptive aspects an empirical ethics can be meaningful. This does not imply that there is not a distinction to be made between descriptive and prescriptive aspects. Indeed, the distinction may very well have its function within a particular context. The point is that making the distinction into an irreducible gap between intrinsic meanings is erroneous [10].

Experience

Thirdly, in its attempt to answer moral questions, empirical ethics defends the belief that those answers must be found empirically, that we must look for answers by means of experience. In empirical ethics experience is seen not only as the source of moral practice but also as the source of ethical theory [11]. Of course we do not assume that empirical research simply *reveals* facts, and that experience is merely the material for reflection. On the contrary, we start from the belief that experience is pregnant with conclusions and reflection. For instance, it is clearly both uninteresting and even impossible to 'simply' observe. All meaningful observation requires an end that we have in view. Only that end decides what is interesting, only the end decides what the relevant facts are. Far from preceding observation, facts are created in virtue of some observation which always involves reflection.

Empirical ethical approaches must defend themselves against a number of widespread conceptions according to which empirical ethics is not possible. One of the objections that may be raised is that the very nature of morality makes the empirical method unsuitable for tackling moral problems. In a nutshell, the criticism amounted to saying that moral judgements are not amenable to the empirical method, because moral judgements are fundamentally different from empirical statements.

So, if we mean business with an integrated empirical ethics, this will have consequences for the meaning of 'the empirical' and 'ethics', respectively.

Two Examples of Empirical Ethics

Empirical approaches actively try to find alternatives to overcome the aforementioned problems. Such a search is inspired by the different, though related views of, respectively, John Dewey's (1859-1952) pragmatic work, Hans-Georg Gadamer's (* 1900) hermeneutics and John Rawls' (* 1921) reflective equilibrium [12].
In order to give a tentative answer to the question whether a turn to experience can achieve the proposed integration of, respectively, theory and practice, and descriptive and

normative elements, we give an example of reflective equilibrium and an example of pragmatic hermeneutics.

Reflective Equilibrium

The study *Respect in nursing home care* started from the observation that the understanding, prevailing in the Netherlands, of the principle of respect for patient autonomy is problematic when it comes to the care of patients who are not fully competent, such as many nursing home residents. The aim of this study was to formulate a so-called 'modest theory' of respect for patient autonomy in nursing home care. Rules for careful decision-making, for instance in a guideline for Do-Not-Resuscitate decisions, may serve as another example of such a modest theory [13]. A specific form of reflective equilibrium was used, namely the *network model* as described by Van Willigenburg and Heeger and enhanced by Van der Burg [14]. The network of considerations used in this case consists of (empirically found) moral intuitions of caregivers, moral principles and ideals. The fact that empirical research attempts to obtain information regarding moral intuitions of relevant agents – other than the researcher, who is looking for reflective equilibrium – makes this study an example of empirical ethical research [15].

Following the categories of four well-known interpretations of respect for patient autonomy in the literature, four views on patient autonomy in the nursing home were described. These views were put into a questionnaire and presented to 100 nurses and 50 nursing home physicians. They were asked to indicate which of these views they would prefer for their own nursing home. The response to this question gives information about the more abstract, general view of caregivers on the subject of patient autonomy.

Apart from the moral view on care in general, the moral intuitions and preferences of caregivers in concrete situations were considered to be relevant. Do caregivers stick to their intuitions about the best way to interpret respect for patient autonomy when confronted with specific aspects of a case? To answer this question ten vignettes were designed. The variables of the vignettes were considered relevant to the question of how respect for patient autonomy should be interpreted in practice.

Following each vignette there were four comments on the case. Each comment was formulated from the perspective of one of the views on respect for patient autonomy and it contained a suggestion for the best way to deal with the given situation. The respondents were instructed to choose the option that best reflected their opinion about the case as opposed to choosing the comment that best described daily practice.

The results of this empirical study [16] showed that 39% of caregivers preferred the libertarian view as a general description of respect for patient autonomy. Another 33% preferred the narrative understanding, 18% the Kantian and 10% the Ethic-of-care perspective. The choice for a view on good nursing home care did not, however, correlate with the choice of approach when confronted with vignettes. In concrete case descriptions, the influence of relevant circumstances (the variables of the case) was significant. On the basis of these results the conclusion was that caregivers tend to shape their respect for autonomy according to specific needs or abilities of patients. They do not think that respect for patient autonomy requires that patients are treated in the same way regardless of the circumstances.

After obtaining empirical data on the moral intuitions of our respondents the further goal was to describe a normative view on respect for patient autonomy in the specific context of the nursing home. In trying to describe a coherent view, the notions that were central to the four views on autonomy were elaborated. The empirical data did not point to one of the four approaches as the most favoured under all circumstances in this context. Instead, what seemed to be needed was a refinement of these approaches fitting the context of the nursing home. This refinement was achieved by a so-called narrow reflective equilibrium: a coherent set of statements based on intuitions and different interpretations of the principle of respect for autonomy of nursing home residents [17]. Clearly, a critical input was necessary to come to a wide reflective equilibrium. This input was found in ideals [18]. These concerned views on the patient-caregiver relationship, ideals concerning the care of a good caregiver and the ideal of autonomous agency itself. The ideals supported some of the aspects of the narrow reflective equilibrium and did not support others. Further normative argumentation about the power and the relevance of the moral intuitions, principles and ideals led to a balanced view on how to respect autonomy in a setting where this cannot be taken for granted for many reasons. This resulted in a concrete proposition for a normative view on respect for patient autonomy that can also be used in policy documents and in instruments designed to define or evaluate quality of care.

Pragmatic Hermeneutics

Pragmatic hermeneutics is a combination of pragmatism and hermeneutics. John Dewey and Hans Georg Gadamer are, respectively, the philosophers who form the source of inspiration of this philosophical approach. Pragmatic hermeneutic insights appear to be particularly suited to fruitfully approaching the problems confronting the idea of empirical ethics. The special significance of pragmatism is in its theory of meaning. In its broadest sense pragmatism is the view that the meaning of a proposition, a belief or a theory, derives from the practical consequences which originate in the usage of such a proposition, belief or theory. Pragmatists hold that the proper task of concepts or theories is not to give an adequate description of a fixed and given reality which would be independent from or prior to such description. On the contrary, starting from the premise that reality is continually changing, they suggest that concepts as well as theories should be seen as means that can be used in the analysis of situations in which we want to influence the course of events. The changes within reality are the effect of interactions. And if we do not want those interactions to be stupidly blind, but as far as possible, to yield predictable consequences, we can make use of concepts and theories. They do not represent any a priori or absolute authority, but they owe their value to their mediation in achieving that for which they were intended to be. Thus, their value is determined by their context.

Hermeneutics in its turn is characterized by the extensive attention it pays to communication and especially to dialogue as a specific form of interaction. The study *Interpretation, action, and communication. Four stories about a supported employment program* [19] gives an illustration of the pragmatic hermeneutic approach of empirical ethics. The study evaluates a government policy concerning persons with a mental handicap. The policy's purpose is to achieve the integration of people with a mental handicap within normal work situations. The mentally handicapped, their parents, a job coach, and the manager on the working place are the people who are directly involved in such management. One way of evaluating the 'supported employment' policy is to investigate whether the ends in view of the proposed policy have been achieved. Relevant questions would be, for instance: Are handicapped people treated as other people are, or are they still perceived and stigmatised as 'different'? Is their individuality enforced and can they make more autonomous decisions regarding their work, or are decisions still made for them by other people? In this method of evaluation, the program of supported employment is measured according to pre-given standards. The idea is that the program can be judged from an independent point of view, and that this judgement tells us whether the policy has been successful or not [20]. This approach makes use of clear distinctions, such as the distinction 'integration' versus 'stigmatisation', or 'autonomy' versus 'paternalism'.

Another way to evaluate the program is to start from the experiences of the people involved. In order to get to know these experiences, one has to listen to the stories of the participants. These stories do not score the programme in terms of pre- given categories; they present concrete judgements about what is good and bad. And these judgments develop in time. Each person involved has his or her own perspective, using different concepts to describe the situation, and evaluating the situation in different ways. These stories, moreover, influence each other. They change the perceptions and actions of the people involved. The investigators who invite the story telling, also have an influence on both the experience of those involved and on the development of the situation. In this approach, the focus is not on isolated individuals initiating certain activities, but on the contrary, on the interactions between the supported employment program, the people involved and the investigators.

The study focused on one case of supported employment (the case of a person called Ben). The stories of the participants in this case show that the results of the supported employment program are complex and unique. All the participants agree that for Ben the program has worked out in a positive way. Yet, one cannot say that the program has resulted in an unambiguous integration. It is clear that Ben is not treated as any other worker; he has to be supported and protected. It is also clear that he is not fully autonomous in the sense of being able to determine his own future. The concept of autonomy that is more fitted to Ben's case is the concept of 'actual autonomy', developed by Agich [21]: Ben identifies with his work and knows his way around. The case makes clear that in this case autonomy is not the same as independence. People with a mental handicap can only become more autonomous by being supported by others.

This evaluation study is ethical in that it aims to make clear what contributes to a good life for people with a mental handicap, and to make explicit the moral responsibilities that are part of the care for the mentally handicapped. The study presupposes that one can learn about the good life and the inherent responsibilities from the experiences of the people involved. The view of the good life and the moral obligations people have towards one another in order to make the good life possible are primarily developed in concrete practices.

The aim of the pragmatic hermeneutic interpretation is to make explicit the normative orientation in daily life. This is not merely a matter of descriptive ethics. Pragmatic hermeneutic interpretation makes use of theoretical and normative orientations themselves. It aims to contribute to a better understanding of what is needed for specific practices to be sustained and improved. In so far as theories are involved (for instance, theoretical notions of autonomy), they are directly related to the practices under consideration (such as the practice of care for the mentally handicapped) . Indeed, the value of terms and theories is evaluated according to their power to do what they were meant to do. In this case the question is whether the notion of autonomy helps to improve Ben's life.

Thus, in the Aristotelian tradition, the aim of the investigation is not the achievement of a universally valid ethical theory, but to find possibilities to improve concrete practical situations, while taking into account as many factors as possible: the situation of the labour market, financial means, the capacities of the handicapped, the requirements of the place of work, etcetera.

Epilogue

We have shown two examples of research within the tradition of, respectively, reflective equilibrium and pragmatic hermeneutics. These examples demonstrate that empirical-ethical research does not maintain a rigid distinction between descriptive and normative issues. Both approaches attempt to do justice to a normativity that is inherent to a specific practice, and to empirically achieve a more adequate form of normativity. While the reflective equilibrium approach focuses specifically on moral intuitions and experiences of caregivers, the pragmatic hermeneutic approach shows an interest in the stories of and about a certain practice. Neither one of these approaches aims at formulating a general ethical theory, but at formulating the normative content of a specific situation. Instead of abstracting, they emphasise precisely that a specific normativity applies to a specific context. Instead of imposing criteria in terms of which the critical function of a normative ethics is fulfilled, without any reference to a context, in reflective equilibrium the criteria are being developed with specific attention for contextual experiences, while in pragmatic hermeneutics they are being developed within the situation itself. Empirical ethics tries to develop normative theories by means of an empirical inquiry into the practice relevant to specific problems at hand.

Our project aims to investigate further the possibilities of empirical ethics. In order to achieve this, we will also make use of empirical methods. We will elaborate the experiences of other scholars who are engaged in empirical-ethical research and enter into a dialogue with them about suitable methods. We specifically focus on research which is actually being performed in the context of the Dutch programme "Ethics and Policy". This programme requires each study to combine empirical and ethical elements. We are interested in the ways in which researchers within this programme relate descriptive and normative issues. We expect to learn from their questions, their problems, their approaches and solutions and we hope that, through our project, we too can make a contribution to new insights that may be profitable for them. Ultimately, our goal is to make a contribution to a qualitative improvement of empirical ethics.

References

[1] As Heeger points out: concreteness is not alien to all previous ethics, and at the present time there is special ethics, but he does state that, influenced by Kant as it was, modern ethics has for the most part been abstract, and that the analyses of moral language that were carried out in Anglo-Saxon philosophy showed no concern for concrete questions. Heeger, R, (1993), What is Meant by 'The Turn to Applied Ethics'? In Heeger, R, and Willigenburg, T, van (Eds.), *The Turn to Applied Ethics. Practical Consequences for Research, Education, and the Role of Ethicists in Public Debate*, Kampen, Kok Pharos, pp.10-11

[2] Beauchamp, TL, and Childress, JF, (1994), *Principles of Biomedical Ethics*, Fourth edition. Oxford, Oxford University Press; Jonsen, AR, and Toulmin, S, (1988) *The Abuse of Casuistry. A History of Moral Reasonin,,.* Berkley, University of California Press

[3] Arnold, Robert M, and Forrow, Lachlan, (1993), "Empirical Research in Medical Ethics: an Introduction", *Theoretical Medicine* 14, p.195

[4] Arnold, Robert M, and Forrow, Lachlan, (1993), op cit, pp.195-6; Birnbacher, D, (1999), "Ethics and Social Science: Which Kind of Co-operation?", *Ethical Theory and Moral Practice*, 4, pp.319-366; Braddock, C, (1994), "The Role of Empirical Research in Medical Ethics: Asking Questions or Answering them?, in *The Journal of Clinical Ethics* 5/2, pp.144-147; Brody, B, (1993), "Assessing Empirical Research in Bioethics", *Theoretical Medicine,* 14, pp.211-219;Crigger, B, (1995), "Bioethnography: Fieldwork in the Lands of Medical Ethics", *Medical Anthropology Quarterly*, 9, pp.400-417; DeVries, R, and Conrad, P, (1998); Hartogh, G den, (1999), "Empirie en Theorievorming in de Ethiek", *K&M: tijdschrift voor empirische ethiek*, 23, pp.172-177; Hoffmaster, B, (1990), "Morality and the Social Sciences", In G. Weisz (ed.) *"Social Science Perspectives on Medical Ethics"* , Amsterdam, Kluwer Academic Publishers, pp.241-260; Hoffmaster, B, (1992), "Can Ethnography Save the Life of Medical Ethics?", *Social Science in Medicine*, 35 /12, pp.1421-1431; Lindemann Nelson, J, (2000), "Moral Teachings from Unexpected

Quarters. Lessons for Bioethics from the Social Sciences and Managed Care", *Hastings Center Report* 30/1, pp.12-17; Musschenga, B, and van der Steen, W, (1999) "Empirie in de Ethiek en Empirische Ethiek", *K&M: Tijdschrift voor Empirische Ethiek*, 23, pp.155-166; Musschenga, AW, (2000), "Empirical Science and Ethical Theory: the Case of Informed Consent", In Musschenga, AW and van der Steen, W, (Eds), *Reasoning in Ethics and Law*, Avebury: Ashgate, pp.183-204; Pellegrino, ED, (1995) "The Limitation of Empirical Research in Ethics", *The Journal of Clinical Ethics*, 6/2, pp.161-162; Widdershoven, (1999), "Ethiek en Empirisch Onderzoek", *K&M: Tijdschrift voor Empirische Ethiek*, 23, pp.145-154; Zussman, R (2000), "The Contributions of Sociology to Medical Ethics", *Hastings center report*, 30/1, pp.7-11

[5] Musschenga, A.W and van der Steen, W (1999), op cit, pp.155-156, 162; Hartogh, G den, (1999), op cit, p.176; Pearlman, Robert A, Miles, Steven H and Arnold, Robert M, (1993), "Contributions of Empirical Research to Medical Ethics", *Theoretical Medicine*, 14, pp.197-210; Brody, Baruch A, (1993), "Assessing Empirical Research in Bioethics", in: *Theoretical Medicine*, 14, pp.211-219, here p. 218; Braddock, C, (1994) op cit, pp.144-147

[6] Hoffmaster, Barry, (1992), op cit, p.1421; Pellegrino, ED, (1995), op cit, p.161-162,

[7] Empirical ethics is called 'new' here, even though we realize that the name empirical ethics was also used by philosophers and ethicists in the period around 1940/1950

[8] ten Have, Henk A M J, and Lelie, Annique, (1998) "Medical Ethics Research between Theory and Practice", *Theoretical Medicine and Bioethics*, 19, pp.263-276; Lelie, Annique, (1999), *Ethiek en Nefrologie. Een Empirisch-Ethisch Onderzoek*, Best, Damon; van Delden, Hans (1999), "De Hherontdekking van de Praktijk", *K&M, Tijdschrift voor Empirische Ethiek*, 23, pp.167-171; Widdershoven, Guy, (2000), "Empirische Ethiek: Op Sleeptouw bij de Medische Technologie?", *Tijdschrift voor geneeskunde en ethiek*, (10) 3, pp.77-82

[9] Pels, Dick and de Vries, Gerard, (1990), "Feiten en Waarden: de Constructie van een Onderscheid", *Kennis en Methode*, 1, pp.7-13, Pels, Dick, (1990), "De 'natuurlijke saamhorigheid' van feiten en waarden", *Kennis en Methode*, 1, pp.14-43

[10] One may object that it is simply impossible to abandon the distinction between descriptive and prescriptive aspects, because that distinction is indisputable. Yet, the insight that theoretical concepts always have their origin in a certain historical constellation and very specific goals, implicitly involves the insight that concepts which appear to refer to evident facts must also have their origin in a historical and therefore more or less contingent setting. So, though the distinction between facts and values is generally considered to be the expression of a philosophical evolution leading from naïve monism to critical dualism, the outcome of this evolution has in fact almost universally come to be regarded as a self-evident timelessly essential distinction.

[11] van der Scheer, Lieke, (1999) *Ongeregelde Moraal. Dewey's Ervaringsbegrip Als Basis Voor een Nieuwe Gezondheidsethiek*, Nijmegen, Valkhofpers

[12] Dewey, J, (1925), *Experience and Nature*, The Later Works 1, Jo Ann Boydston (Ed.) London [etc.] Illinois University Press; Gadamer, H G, (1960) *Wahrheit und Methode, , Grundzüge einer philosophische Hermeneutik*, Tübingen; Mohr, JCB, (1960); Rawls, J, (1972), *A Theory of Justice*, Oxford, Oxford University Press

[13] van Delden, JJM, (1993), *Beslissen om Niet te Reanimeren. Een Medisch en Ethisch Vraagstuk*. Assen: Van Gorcum, 1993

[14] Van Willigenburg, T and Heeger, FR, (1989), Justification of Moral Judgements: A Network Model. In: *Societas Ethica Jahresbericht 1989*. Hannover, Societas Ethica, pp.53-61; van der Burg, W, (1991), *Het Democratisch Perspectief: Een Verkenning van de Normatieve Grondslagen der Democratie*, Arnhem, Gouda Quint

[15] van Thiel, GJMW, and van Delden, JJM, (2001), The Principle of Respect for Patient Autonomy in the Care of Nursing Home Residents, *Nursing Ethics*.8, nr 5, pp.419-31; van Delden, JJM, and van Thiel, GJMW, (1998), Reflective Equilibrium as a Normative-Empirical Model in Bioethics, In van der Burg, W, and van Willigenburg, T (Eds), *Reflective Equilibrium*, Deventer Kluwer, 1998, pp.251-259

[16] The response among nurses was 94% and 62% of the physicians participated in the study.

[17] DePaul, MR, (1993), *Balance and Refinement. Beyond Coherence Methods of MoralInquiry*. London/New York, Routledge, p.18-19; van der Burg, W, (1991), *Het Democratisch Perspectief. Een Verkenning van de Normatieve Grondslagen der Democratie*, Arnhem, Gouda Quint, p.20

[18] van der Burg, W, (1991), op cit, p.23;.and van der Burg W, (1997), The Importance of Ideals. *The Journal of Value Inquiry* 31 p.26

[19] Widdershoven, Guy and Sohl, Carlo, (1999), "Interpretation, Action, and Communication. Four Stories about a Supported Employment Program", *Advances in Program Evaluation*, 6, pp.109-130

[20] Widdershoven, Guy and Sohl, Carlo, (1999) op cit, p.110

[21] Agich, GJ, (1993), *Autonomy and Long-Term Care*, Oxford, Oxford University Press

Section 3

Empirical Bioethics in Practice

An Empirical Study of the Informed Consent Process of a Clinical Trial*

Anne GAMMELGAARD
Department of Medical Philosophy
University of Copenhagen

Introduction

This paper shows how an empirical study of the informed consent process of a clinical trial may provide for an ethical analysis of the question of whether or not informed consent should be sought in trials involving this particular patient population.

The empirical study concerned the informed consent process of a recent Danish randomised multi-centre trial, the DANAMI-2 study (second DANish Acute Myocardial Infarction study) [1]. This clinical trial, comparing an interventional approach (primary angioplasty) with a medical strategy (fibrinolysis), involved patients who suffered an acute myocardial infarction (AMI) and who were admitted to the emergency department in the acute phase of the disease. The primary angioplasty was only performed in 5 (the angioplasty centres) of the 29 participating hospitals. Patients admitted to the remaining 24 hospitals (the referral hospitals) were randomised to immediate treatment with fibrinolytic therapy in the local hospital or acute ambulance transfer to an angioplasty centre for primary angioplasty. The informed consent process took place while the patient was still on the ambulance-stretcher. Since AMI patients are to be medically treated as soon as possible, the physician only had a short period of time (ideally less than 20 minutes) in which to decide whether the patient was eligible for the DANAMI-2 trial, inform the patient, obtain a consent from the patient, randomise, and initiate the therapy. The informed consent process included oral as well as written information (a 1-page sheet) and patients had to sign a consent form.

From an ethical perspective, this study was extremely interesting since it involved some possible dilemmas related to the informed consent requirements in clinical research. The basic requirements of informed consent are that (a) patients must be informed about the research project and about alternative therapeutic options, (b) patients must be competent, i.e. be able to understand the information and to make a decision, and (c) patients must be allowed to decide for themselves whether or not to participate [2]. Ever since the Danish Parliament's passing legislation on Research Ethics Committees (Law No.503 of 24th June 1992, changed by Law No.499 of 12th June 1996), informed consent from research participants has been mandatory unless they are minors, unconscious, mentally deficient, demented, psychotic, or in other ways obviously incompetent (in which case the consent is to be obtained from the closest relatives or a guardian).

Patients eligible for the DANAMI-2 trial, however, arrive at the emergency department in the midst of an acute myocardial infarction and require immediate treatment. Even though they are conscious, they are in a medical condition that may give rise to particular difficulties in relation to the informed consent process. First of all, it is debatable to what extent acute patients with a serious disease are entirely competent to make a decision since they may be in a state of shock or crisis. Secondly, acute patients have to decide quickly whether or not to participate in the trial and have little time in which to obtain information and to consider and discuss the options. Thirdly, under such circumstances the process of informed consent may be distressing to patients, and it is far from self-evident that patients would *want* to make that kind of decision. Finally, the very process of informed consent, which is supposed to protect patients, may be a cause of harm in itself due to the delay in the provision of therapy which it causes.

Accordingly, the ethical dilemma of the DANAMI-2 study can be seen as a strong version of the ethical dilemma facing clinical trials in general. On the one hand, there is the need to protect the individual patient, and on the other hand there is the need to develop new and better treatments for a particular group of patients.

A literature review revealed that a large number of AMI patients have been enrolled in trials in recent years on the basis of more or less comprehensive consent procedures. Little is known, however, about how patients perceive the various enrolment procedures used in these trials [3], and the literature on the specific ethical dilemmas facing such trials was sparse. In consequence, this research project was dedicated to a systematic analysis of how patients involved in the informed consent process of the DANAMI-2 trial perceived this process. The empirical part of the project included a qualitative interview study and a questionnaire based follow-up survey with patients who participated and patients who did not participate in the DANAMI-2 trial. Unlike earlier empirical studies of the consent process in AMI trials, this study was thus to include a substantial number of patients who did not consent to the trial in order to enable us to analyse why they did not give their consent and whether their perception of the consént process differed from that of the trial participants. On the basis of the empirical study, it was an aim 1) to analyse and assess the ethical issues of clinical trials involving AMI patients, 2) to provide guidelines for the consent process of future trials involving AMI patients, and 3) to analyse and assess present Danish informed consent legislation in relation to clinical trials involving AMI patients.

Methods

The empirical study involved both qualitative and quantitative methods as these were intended to supplement each other and ensure that the data would be sufficiently comprehensive [4]. The interviews with patients were expected to provide for a detailed understanding of the relevant ethical issues, and the questionnaire survey was expected to provide for a statistical analysis of factors influencing the informed consent process. Accordingly, it should be possible, on the basis of the qualitative methods, to describe patients' perceptions of the consent process and secondly, it should be possible, on the basis of the survey, to analyse frequency distributions and statistical associations. While the qualitative methods would reveal the character of the ethical issues, the questionnaire survey would estimate the extent of them.

Accordingly, the discussions and conclusions in this paper are based on several methods and various sources of data. The decision to use several methods in this study was based on the assumption that this would provide for a comprehensive and precise answer to the research questions since the methods - each with different strength and weaknesses - would supplement each other.

A Qualitative Study of Patients' Perception of the Informed Consent Process

Semi-structured interviews were conducted with a total of 32 patients who either consented to participation in the trial or did not give consent. This study and its results are fully described elsewhere [5].

In any attempt to draw general conclusions on the basis of a qualitative study, one needs to be cautious. A qualitative study is generally based on a limited number of cases, and in this study the cases were moreover sampled selectively to include as wide a range of informants as possible. In consequence, the study population is not necessarily representative of the total DANAMI-2 patient population.

It is sometimes claimed that this is not a substantial problem since qualitative studies are not *meant* to be generalisable. Mays & Pope explain that such claims originate from extreme relativists who hold that all research perspectives are unique and each is equally valid in its own terms [6]. But it seems unlikely that qualitative researchers including extreme relativists would not, at least to some extent, aim for results that are relevant and applicable outside the particular study setting. If the results of a study were applicable only to the study itself, the relevance of that study to anyone not actually involved in it would be highly questionable.

If, however, the results of qualitative studies like this one cannot be generalized in the statistical sense of the word, what kind of generalisations would then be valid? This important issue has been much discussed [7]. Conrad suggests that generalisability in qualitative research should be understood in terms of *concepts* rather than in terms of *samples*. He argues that the issue of generalisability in relation to, for instance, the concept of *deviance disavowal* described in relation to a group of polio sufferers "has nothing to do with polio or equivalent samples or disabilities, but more to do with how generalisable is the concept *deviance disavowal* to social processes concerning other forms of illness, disability or social life". According to Conrad's account then, it seems that the generalisability of a concept refers to its *applicability* in other contexts. Likewise, Schultz Jųrgensen emphasises what he terms the *analytical generalisability* of qualitative research. According to Schultz Jųrgensen, qualitative research should be understood as a method to develop and evaluate theoretical concepts, and he stresses that the validity of a qualitative study is intimately connected with its methodology. In consequence, the validity of a qualitative study can be assessed by examining how patients were sampled, how the interviews (of an interview study) were conducted, how the analyses were carried out, and how the conclusions were arrived at.

Arguably, a study needs to be valid to justify any attempt to generalise on the basis of it. In this sense, Schultz Jųrgensen's account of validity constitutes a *necessary* condition of generalisations. However, it is far from evident how it could possibly constitute a *sufficient* condition. If, for instance, we have developed and refined a concept by carrying out a qualitative study of high validity, it seems that we would still be in no position to claim that this concept would be applicable in another context or sample. In consequence, it is difficult to see how the notion of *analytical generalisability* solves this basic problem of qualitative studies. It seems that the problem remains the same whether we wish to generalise from one sample to another or to apply a concept derived from one context to another.

It seems more reasonable to recognise that the difficulties connected with the issue of generalisability are essential limitations of qualitative studies. This recognition, however, does not imply that no generalisations from qualitative studies are ever permissible. Rather, my point is that, compared to studies based on representative samples, qualitative researchers need to consider carefully what kind of generalisations may be appropriate on the basis of a particular study, and they need to give *reasons* for the tenability of each generalisation. Such considerations would involve the sampling procedure, the context of the particular study, and the results from the study.

In what sense then are the results from the qualitative study of the DANAMI-2 trial generalisable? On the basis of the study sample, I would not be able to predict, for instance, how many patients would be likely to understand the basic information, feel competent, or be satisfied with the consent process in the total population of DANAMI-2 patients. To be able to make such predictions, I would need the results of a survey study (a study based on a representative sample). However, I would assume that the kind of perceptions described in the study are likely to be found among the rest of the DANAMI-2 patients too, because the patients were sampled deliberately to provide for descriptions of a wide variety of perceptions.

Furthermore, I would assume that similar kinds of perceptions would be found among patients participating in future AMI trials, unless those trials, the patients, or the circumstances *deviate* from those of the DANAMI-2 trial with regard to aspects of relevance to the informed consent process.

A Questionnaire-Based Survey of Patients' Perception of the Consent Process

To ensure that the qualitative interview study and the questionnaire survey would supplement each other, the questionnaires of the survey were designed on the basis of the interview study. Two questionnaires (A and B) were developed; one questionnaire (A) was designed for those patients who chose to participate in the DANAMI-2 trial, while the other (B) was designed for those patients who did not wish to participate in the DANAMI-2 trial or who were not capable of consenting. The survey was made as a questionnaire-based follow-up study of the informed consent process of the DANAMI-2 trial. Questionnaire A was mailed to 125 consecutive patients who did participate in the DANAMI-2 trial three weeks after the informed consent process. Questionnaire B was mailed to 122 consecutive patients who did not participate in the DANAMI-2 trial. Most of the items in the questionnaires concern patients' attitudes, perceptions and experiences. Such issues are notoriously difficult to explore by means of a questionnaire with fixed answer categories since many nuances and complexities cannot be taken into consideration. Since those nuances and complexities are described in the interview study, however, the aim of this survey was to supplement that study with some statistical analyses. This study and its results are fully described elsewhere [8].

Implications of the Empirical Study

In the debate concerning the ethics of seeking informed consent in AMI trials, several arguments against informed consent have been presented. First of all, some sceptics have argued that informed consent makes no sense in a situation in which patients' abilities to make autonomous decisions are severely affected by their medical condition and time limits [9]. Secondly, it has been claimed that the very process of informed consent may potentially harm AMI patients since the provision of therapy is delayed until the patient has made the decision [10]. Thirdly, some have argued that AMI patients do not *want* to make such decisions at a moment of crisis [11]. Finally, Truog et al. have argued that there is *no need* to seek informed consent if the trial fulfils the following requirements: 1) the trial should compare two therapies that are already in use; 2) the treatments should not involve more than minimal additional risk in comparison with any of the alternatives; 3) genuine equipoise (professional disagreement over which treatment would be most beneficial to the patients) should exist and 4) no reasonable person should have a preference for one
treatment over the other(s) [12]. The DANAMI-2 trial does not fulfil the fourth criterion since the empirical study shows that some patients had personal preferences for either the invasive or the medical treatment, and it is also debatable whether the trial fulfils the second criterion insofar as patients admitted to the referral hospitals are concerned. When those patients were allocated to the primary angioplasty they had to be transported, and it was not obvious that this transportation involved only minimal additional risk. The first and the third criterion, however,

were fulfilled when the DANAMI-2 trial was initiated. While fibrinolysis was standard treatment in most Danish hospitals, the Skejby University Hospital and the Ålborg Hospital had replaced the fibrinolysis with the primary angioplasty during daytime when specialists able to perform the primary angioplasty were at hand. This situation illustrates a professional disagreement over the evidence from earlier trials i.e. a state of genuine equipoise.

Arguably, if Truog et al's criteria are fulfilled, there are reasons to believe that from a medical perspective, it would not be against patients' best interest to participate in the trial since all treatments offered in the trial are 'state of the art'. This is a situation which challenges the orthodoxy that a clinical trial will *only* benefit future patients at the expense of the patients participating in the trial; an orthodoxy that has also been under attack from proponents of evidence-based medicine [13]. The medical fact that a trial is not against patients' best interest, however, does *not* imply that informed consent is not needed in this trial. Truog et al. fail to show how their conclusion that informed consent is not needed follows from the four criteria.

On the basis of the empirical study of the informed consent process of the DANAMI-2 study, I will argue that informed consent should be sought in such trials despite the possible inconveniences of this process. The basic arguments in favour of informed consent are that: 1) the majority of patients found that they retained their ability to decide and 2) only few patients would want to be included in the trial without giving their consent.

Many patients, however, were well aware that autonomous decision-making (in the sense of reaching a well-informed decision after due deliberation) was not exactly what they were practising at the time of the consent, and most patients were not aware of the possible risks of the trial. For informed consent to make sense in the emergency situation, then, its aim must be somewhat different from that of enhancing autonomous decision-making and protecting patients against research related harms. In her recent book, *Autonomy and Trust in Bioethics*, O'Neill argues that the requirement of informed consent should not be grounded in one or another conception of individual autonomy such as liberty; independence; self-assertion; freedom from obligation; absence of external causation; self-knowledge; critical reflection, or knowledge of one's own interest [14]. Instead, she argues that informed consent requirements are justified because they have the potential to prevent and limit deception and coercion of patients. In consequence, she finds it more revealing to reverse the direction of argument, suggesting that conceptions of individual autonomy may gain such weight as they have from their link to informed consent requirements, which are justified by their contribution to preventing and limiting deception and coercion of patients and relatives [15]. O'Neill also sees informed consent important as a measure to preserve a basis of trust in the relationship between patients and physicians. This modest, but nevertheless important, role of informed consent makes sense particularly in the context of AMI trials. It follows from this account of the role of informed consent that even if the consent is only relatively informed or does little to protect the patients from research related harms, it would be ill-advised to waive it altogether. Likewise, Ågard, Hermeren & Herlitz [16] suggest that what really seems to matter is not that patients are fully informed, but that their right to say yes or no is respected. Despite the difficulties and inconveniences of the process, the AMI patients participating in this study generally found the informed consent process important.

The quality of the informed consent process also seemed to be of considerable importance to patients who participated in it. Patients who felt uninformed, unable to decide, or coerced at the time of the consent felt uncomfortable and dissatisfied with the informed consent process. Accordingly, the requirements of informed consent, as defined by Faden & Beauchamp [17], are relevant not only from a professional bioethical perspective but also from a patient perspective.

Concerning patients' understanding of the information, previous empirical studies of the consent process in AMI trials support this study in the sense that a certain proportion of the patients in those studies seemed relatively informed while a substantial minority did not understand the basic information [18].

Concerning the issue of competence, however, there seems to be no consensus as to whether or not AMI patients are able to decide at the time of the consent. Some studies question patients' abilities [19] while others are more optimistic [20]. Smithline, who found that at least 68% of patients were able to give an informed consent, is supported by our survey study in which 76% of the participants agreed or mostly agreed that they felt able to make the decision. The divergent results in the various studies in relation to patients' competence may reflect an underlying disagreement over the type of cognitive skills that patients must demonstrate to give a valid consent. In our survey study, we focused on patients' self-assessed ability to decide, and we demonstrated that this factor had a substantial influence on the quality of the consent.

In the recent HERO-2 consent substudy of the consent process of an AMI trial, 52% of the patients had the ability to make a treatment choice, 17% also understood the consequences of their choice, and 13% gave rational reasons for their choice [21]. Eighteen percent of the patients, however, were unaware of the consent process. The investigators concluded that most of the patients (52%) only met a very basic autonomy criterion, i.e. being able to do something rather than nothing. In consequence, this empirical study illustrates the relevance of an O'Neillian justification of the informed consent requirements in AMI trials.

Proposals for Improving the Informed Consent Process

In accordance with previous studies, the questionnaire survey and the qualitative study both showed that the oral information was the most important source of information since few patients read the written information sheet. Consequently, a careful review of the oral information by the Research Ethical Committee should be required in AMI trials.

I would suggest that physicians involved in the informed consent process should be supervised about the aim, scope, and difficulties of this process to ensure that patients are not coerced into participating in the trial. If a patient does not feel able to decide, the physician needs to ask whether the patient needs more information or would prefer not to participate in an informed consent process under the circumstances. Patients who are not able to decide or who do not want to decide whether or not to participate in the trial should not be included in the trial.

Further Protection of Patients

The empirical studies show that the informed consent procedures do little to protect patients from research related harms since few patients were aware of the potential risks involved in participating in the trials. Accordingly, patients in general and AMI patients in particular must be protected by other means. The assessment of whether or not it is safe to join a clinical trial should rest with an independent Data and Safety Monitoring Board (DSMB) responsible for overseeing the trial and performing interim analyses. The DSMB of the DANAMI-2 trial (the Safety and Ethical Committee) played a prominent role in relation to the decision to terminate the trial. While it is customary for multi-centre trials to establish such committees, the very existence of a DSMB is no guarantee, however, that a trial will be terminated in due time without unnecessary casualties or harm to patients allocated to the 'wrong' arm of the trial. In a recent paper, Freeman et al. describe how the lethal effect of one treatment in a trial was not detected by the DSMB because they, like the investigators, were blinded to the patients' treatment assignments [22]. Since the overall mortality rate in that trial did not differ significantly from the expected mortality rate, the fact that the mortality was much lower in one arm of the trial than in the other was not detected until the treatments groups were un-blinded. To avoid similar situations in future trials, it is crucial that the DSMBs access to the data is *unrestricted* like in the DANAMI-2 trial. Otherwise, there is simply no point in having a DSMB.

Apart from an independent and unrestricted DSMB, an efficient Research Ethics Committee is crucial to protect patients from research related harms. Due to the emergency circumstances and the limitations of the informed consent process, AMI research protocols should be reviewed extremely carefully. Patients eligible for AMI trials should not be subjected to more than minimal risks from the research related procedures, i.e. the *non-therapeutic* procedures of the trial. The non-therapeutic procedures in the DANAMI-2 trial were minimal risk procedures like, for instance, extra blood samples and check-ups (exercise tests and ECG evaluations) as well as repeated quality of life assessments by means of questionnaires. Since many AMI patients are likely to be unaware of the risks involved in participating in the trial, it would be problematic from a moral point of view to subject those patients to procedures which are riskier than routine physical examinations or tests.

It should be stressed that this minimal risk requirement is not intended to be applied to interventions or procedures that could be beneficial to the patients, i.e. the *therapeutic* procedures of the trial. The therapeutic procedures of the trial are to be justified by clinical equipoise [23]. A novel treatment may pose more risk than standard treatment but offer the prospect of greater benefit. Clinical equipoise requires rough equality in the therapeutic index, a compendious measure of the potential benefits and adverse effects, among the treatments in the trial. Although this distinction between the therapeutic and non-therapeutic elements of a trial is not always fully acknowledged, it is arguably essential in relation to the risk assessment of a clinical trial [24].

The Danish Research Legislation

The Danish law on Research Ethics Committees (Law No.503 of 24th June 1992, changed by Law No.499 of 12th June 1996) is currently under revision. It is the aim of this revision to incorporate the Council of Europe's Convention on Human Rights and Biomedicine, the Guideline for Good Clinical Practice adopted by the EU in 2001 (2001/20/EF), as well as the 5th version of the Helsinki Declaration into Danish legislation. Accordingly, a white paper was circulated in the autumn 2002, following which a revised bill was introduced by the Minister for Research and Information Technology on December 5th, 2002 [25].

In its initial white paper, the government suggested that it should be possible to waive the informed consent in connection with trials involving unconscious or semi-conscious emergency patients. As for the semi-conscious emergency patients, however, this suggestion is highly problematic. First of all, the term *semi-conscious* is rather vague and it is far from clear which patients would be included in this category. For instance, would it include all AMI patient or maybe just some of them? To ensure that a waiver of informed consent is used only when it is absolutely necessary, the conditions of it must be specific and transparent. Secondly, on the basis of patients' perceptions of the DANAMI-2 trial, I concluded that informed consent *should* be sought from such patients.

The suggestion to allow for a waiver of informed consent in trials involving semi-conscious emergency patients, however, did not pass the hearing and so does not figure in the bill introduced in Parliament last December. Consequently, the waiver of informed consent included in this bill only includes emergency patients who are temporarily but definitely incompetent such as unconscious patients. Accordingly, if this bill is approved by the Danish

Parliament, informed consent would still have to be sought in AMI trials.

Nonetheless, certain requirements concerning trials involving emergency patients who suffer a life-threatening disease would still improve the bill. First of all, an independent Data and Safety Monitoring Board should be required in such trials. Secondly, it should be emphasised in the bill that emergency patients who are not unconscious should be able to give an informed consent on the basis of the oral information alone. Accordingly, a Research Ethics Committee review of the oral information should be required. On the other hand, there is no need to abolish the requirement of a written information sheet since the empirical studies show that many patients would like to read it after having recovered from the critical phase. The latest version of the Helsinki Declaration does not require that patients' consent is obtained in writing: "If the consent cannot be obtained in writing, the non-written consent must be formally documented and witnessed"[26]. Since it is debatable whether the requirement of a written consent in the Danish research legislation primarily serves the patients' interest or as a measure to protect physicians from legal liability, this requirement should perhaps be waived in AMI trials. However, the requirement of a written consent should only be waived if such a waiver would benefit the patients and if the oral consent is duly documented and witnessed.

Conclusions

On the basis of the empirical study of the informed consent process of the DANAMI-2 trial, it is concluded that:
- Informed consent should be sought in trials involving AMI patients.
- The informed consent process should mainly be based on oral information
- The Research Ethics Committee should be required to review the oral information.
- The primary aim of the informed consent process is to uphold patients' right to decide whether or not to be enrolled in research.
- Physicians involved in the informed consent process of AMI trials should be supervised about the aim, scope and difficulties of the informed consent process.
- Patients' ability to decide should be addressed during the informed consent process and only patients who retain their ability to give an informed consent should be included in research.

Despite the fact that most AMI patients seem able to give an informed consent, they are still a vulnerable population and so need additional protection in relation to research.
- Research protocols involving AMI patients should be reviewed extremely carefully by the Research Ethics Committee.
- Research related procedures should subject patients to no more than minimal risks.
- An independent Data and Safety Monitoring Board with unrestricted access to the data should be mandatory in such trials.

*This paper is based on the introductory chapter of my PhD Thesis (Gammelgaard,A. (2003). *Ethical Aspects of Clinical Trials involving Acute Patients - described in relation to the DANAMI-2 trial*. PhD Thesis. University of Copenhagen), and was partly presented at the EMPIRE Project First Plenary Meeting in Copenhagen, 15-18 June 2000

Acknowledgements

I would like to thank the patients and relatives who participated in the study. I would also like to thank the DANAMI-2 steering committee, in particular Henning Rud Andersen, Peer Grande and Torsten Toftegaard Nielsen for making this study possible and facilitating the collaboration with the hospitals concerned. Furthermore,

I would like to thank Niels Gadsbųll, Anne Rahbek Thomassen, Hans Ibsen, clinicians and study nurses for providing helpful information on DANAMI-2. I am grateful to the Departments of Cardiology at the hospitals participating in the DANAMI-2 for enabling me to contact the patients. I would also like to thank Hanne Andersen, Stig Brorson, Flemming Binderup Gammelgaard, Ole Steen Mortensen, Michael Norup and Peter Rossel for reviewing and commenting on the manuscript. This work was supported by the Faculty of Health Sciences, University of Copenhagen and Eva & Henry Frenkels Memorial Fund. The DANAMI-2 trial was supported by the Danish Heart Foundation, The Danish Medical Research Council, Astra-Zeneca, Bristol-Myers Squibb, Cordis a Johnson & Johnson Company, Pfizer, Pharmacia-Upjohn, Boehringer Ingelheim and Logic I/S Guerbet SA.

References

[1] Andersen, HR, Nielsen,TT, Rasmussen, K, Thuesen, L, Kelbaek, H, Thayssen, P, Abildgaard, U, Pedersen F, Madsen, JK, Grande,P, Villadsen, AB, Krusell, LR, Haghfelt,T, Lomholt, P, Husted, SE, Vigholt,E, Kjaergard, HK, & Spange Mortensen, L (In review). Acute Coronary Angioplasty Versus Fibrinolytic Therapy in Acute Myocardial Infarction: the Danish Multicenter Randomized Trial (DANAMI-2), *The New England Journal of Medicine*

[2] Faden, RR, and Beauchamp,TL, (1986), *A History and Theory of Informed Consent,* Oxford: Oxford University Press

[3] Gammelgaard, A, (In press), Informed Consent in Acute Myocardial Infarction Research, *Journal of Medicine and Philosophy*

[4] Barbour, RS, (1999), The Case for Combining Qualitative and Quantitative Approaches in Health Services Research, *Journal of Health Services Research & Policy*, 4, 39-43; Denzin, NK, (1978), *Sociological Methods* New York, McGraw-Hill; Holstein, B, (1996), Triangulering - Metoderedskab og Validitetsinstrument. In Lunde, IM, and Ramhøj, P, (Eds.), *Humanistisk forskning Indenfor Sundhedsvidenskab*, København: Akademisk Forlag; Mays, N, and Pope, C, (2000), Qualitative Research in Health Care. Assessing Quality in Qualitative Research, *BMJ*, 320, 50-52

[5] Gammelgaard, A, Rossel, P, & Mortensen, OS, (Accepted for publication), Patients' Perspectives of Informed Consent in Acute Myocardial Infarction Research: a Study of the Consent Process in the DANAMI-2 Trial, *Social Science and Medicine*

[6] Mays, N & Pope, C, (2000), op cit

[7] Conrad, P, (1990), Qualitative Research on Chronic Illness: a Commentary on Method and Conceptual Development, *Social Science & Medicine*, 30, 1257-1263; Kvale, S, (1996), *Interviews: An Introduction to Qualitative Research Interviewing,* London: Sage; Malterud, K, (1996), *Kvalitative Metoder i Medisinsk Forskning,* Tano Aschehoug; Mays, N, and Pope, C, (1995), Rigour and Qualitative Research, *BMJ*, 311, 109-112; Mays, N & Pope, C, (2000), op cit; Schultz Jørgensen, P, (1996), Generalisering - i Kvalitativ Forskning, In Lunde, IM, and Ramhøj, P, (Eds.), *HumanistiskFforskning Indenfor Sundhedsvidenskab,* København: Akademisk Forlag

[8] Gammelgaard, A, Mortensen, OS, and Rossel, P, (In review), Patients' Perceptions of Informed Consent in Acute Myocardial Infarction Research: a Questionnaire Based Survey of the Consent Process in the DANAMI-2 Trial

[9] Rasmussen, HH, Hansen PS, Koyama, Y, Adelstein, BA, O'Connell, AJ, and Nelson, GI, (2001), Trial of a Trial by Media, *Medical Journal of Australia*, 175, 625-628; Smith, HL, (1974), Myocardial Infarction - Case Studies of Ethics in the Consent Situation, *Social Science & Medicine*, 8, 399-404, Verdu-Pascual, F and Castello-Ponce, A, (2002), Informed Consent Doesn't Exist in AMI Trials, *Journal of Medical Ethics*, 28, 190-191

[10] Collins, R, Doll, R, and Peto, R, (1992), Ethics of Clinical Trials, In Williams, CJ, (Ed.), *Introducing New Treatments for Cancer Practical, Ethical and Legal Problems,* Chichester: John Wiley & Sons Ltd, pp. 49-65; Tognoni, G, and Geraci, E, (1997), Approaches to Informed Consent, *Controlled Clinical Trials*, 18, 621-627

[11] Tobias, JS, (1997), BMJ's Present Policy (Sometimes Approving Research in which Patients Have Not Given Fully Informed Consent) Is Wholly Correct. *BMJ*, 314, 1111-1114; Tognoni, G, and Geraci, E, (1997), op cit

[12] Truog, RD, Robinson, W, Randolph, A, and Morris, A, (1999), Is Informed Consent Always Necessary for Randomized, Controlled Trials? *New England Journal of Medicine*, 340, 804-807

[13] Ashcroft, R, (2000), Giving Medicine a Fair Trial. Trials Should Not Second Guess What Patients Want, *BMJ*, 320, 1686; Chalmers, I, (1995), What Do I Want From Health Research and Researchers When I Am a Patient? *BMJ*, 310, 1315-1318

[14] O'Neill, O, (2002), *Autonomy and trust in Bioethics,* Cambridge: Cambridge University Press

[15] O'Neill, O, (2002), ibid

[16] Ågard, A, Hermeren, G, and Herlitz, J, (2001), Patients' Experiences of Intervention Trials on the Treatment of Myocardial Infarction: Is it Time to Adjust the Informed Consent Procedure to the Patient's Capacity? *Heart*,

86, 632-637

[17] Faden, RR, and Beauchamp,TL, (1986), op cit

[18] Kucia, AM, and Horowitz, JD, (2000), Is Informed Consent to Clinical Trials an "Upside Selective" Process in Acute Coronary Syndromes? *American Heart Journal*, 140, 94-97; Ockene, IS, Miner, J, Shannon, TA, Gore, JM, Weiner, BH, and Ball, SP, (1991), The Consent Process in the Thrombolysis in Myocardial Infarction (TIMI-- phase I) Trial, *Clinical Research*, 39, 13-17; Smith, HL, (1974), op cit; Williams, BF, French, JK, and White, HD, (1997), Is Our Method of Obtaining Consent Appropriate for Randomised Controlled Trials in Acute Myocardial Infarction? *New Zealand Medical Journal*, 110, 298-299; Williams, BF, French, JK, and White, HD, (2003), Informed Consent During the Clinical Emergency of Acute Myocardial Infarction (HERO-2 consent substudy): a Prospective Observational Study, *The Lancet*, 361, 918-922; Yuval, R, Halon, DA, Merdler, A, Khader, N, Karkabi, B, Uziel, K, and Lewis, BS, (2000), Patient Comprehension and Reaction to Participating in a Double-Blind Randomized Clinical Trial (ISIS-4) in Acute Myocardial Infarction, *Archives of Internal Medicine*, 160, 1142-1146; Ågard, A, Hermeren, G, and Herlitz, J, (2001) op cit

[19] Smith, HL, (1974), op cit; Ågard, A, Hermeren, G, and Herlitz, J, (2001) op cit

[20] Smithline, HA, Mader, TJ, and Crenshaw, BJ, (1999), Do Patients with Acute Medical Conditions Have the Capacity to Give Informed Consent for Emergency Medicine Research? *Academic Emergency Medicine*, 6, 776-780

[21] Williams, BF, French, JK, and White, HD, (2003), op cit

[22] Freeman, BD, Danner, RL, Banks, SM, and Natanson, C, (2001), Safeguarding Patients in Clinical Trials with High Mortality Rates, *American Journal of Respiratory and Critical Care Medicine*, 164, 190-192

[23] Freedman, B, (1987), Equipoise and the Ethics of Clinical Research, *New England Journal of Medicine*, 317, 141-145

[24] Weijer, C, (1999), Thinking Clearly about Research Risk: Implications of the Work of Benjamin Freedman, *IRB*, 21, 1-5

[25] Forslag til Lov om et videnskabsetisk komitesystem og behandling af biomedicinske forskningsprojekter [The research ethics committees and approval of biomedical research projects bill]. 5th December 2002

[26] Helsinki Declaration, 2000, §22

Stimulating Public Debate on the Ethical and Social Issues Raised by the New Genetics

Dr. Mairi LEVITT
University of Central Lancashire

Kate WEINER
University of Nottingham

John GOODACRE
Postgraduate School of Medicine and Health, University of Central Lancashire

With the rapid development of genetic research and applications in health care there is a consensus that some form of public consultation about these new technologies is necessary. Public consultations in the area of medical genetics to date have tended to involve one of two approaches: first, large-scale surveys of opinion providing structured responses to a range of fixed-response questions requiring yes/no answers; second, smaller-scale investigations, such as consensus conferences or focus groups, which provide more in-depth information. Aims vary from public education to a desire for the public to influence the development and regulation of genetic technologies. This paper discusses the results of one public consultation, sponsored by the Wellcome Trust, undertaken with a local daily newspaper in the north-west of England, the Lancashire Evening Post. Illustrated articles were published each day in a 'Gene Week' covering aspects of current interest in medical genetics and including the views of regional medical specialists, local health professionals and patients. The content was designed to inform debate about the social and ethical issues raised by the new genetics. Open-ended questions followed each article and readers were invited to respond with their views by post or through the newspaper's website. This chapter discusses the findings of the Gene Week consultation and the value of stimulating public debate on bioethical issues. The conclusion is that empirical research findings such as these are relevant to the work of bioethicists.

The traditional view of the public understanding of science sees the public as ignorant about science/genetics and their 'deficit' of knowledge as linked to a hostility or lack of support for new developments in genetics. Concern about public ignorance comes from scientists, biotechnology companies and national governments who see it as a barrier to technical progress in genetics and to potential economic opportunities [1]. For example, headlines in UK papers in May quoted the British Prime Minister saying 'the government is not going to allow misguided protests against science to get in the way of confronting the challenges of making the most of our opportunities'- by 'get in the way' he meant drive biotechnology investment to other countries [2]. He was speaking particularly about the anti-GM food movement and attitudes to research using animals. However, research has shown that there is no simple link between ignorance of biotechnology and hostility towards it.

The Link between Knowledge and Attitudes

A review of European surveys of the public found that technical knowledge of biotechnology has increased overall but support for its applications has declined [3]. A UK study found that as people knew more about cloning they came up with more arguments against it [4]. In this study background information was provided on basic genetics and the techniques of cloning. However, the focus group discussions centred on the implications for individuals and society, ethical issues, regulation and the motivation for doing the research. The goal of scientific literacy for the public is assumed to consist of knowledge about scientific facts and technical details. But, as Zimmerman points out, in other contexts literacy is not just a technical matter of being able to read the words but is about understanding, being able to read 'critically and effectively' [5]. Given the opportunity, lay people ask questions about the context in which scientific facts are produced: funding arrangements, the types of research that are supported, the interests of the biotechnology industry and governments as opposed to patients and the general public and about the practicalities of using genetic technology in the current social conditions and health care system [6]. It is these issues that influence people's attitudes towards technologies rather than their ignorance or knowledge of the technical 'facts'.

Why Consult the Public?

Public participation in the process of science and technology policy is a growing phenomenon [7]. There seems to be a consensus that the public should be consulted about new genetic technology and applications in health but there are different rationales for this. First, as already discussed, some still assume that more understanding will lead to greater acceptance of technologies. Second, that consultation will provide more openness and dialogue between science and society leading to a restoration of the basic trust in science and science governance that a modern high technology society requires [8]. In the UK this has been prompted by a number of science policy crises e.g. the handling of BSE (mad cow disease) and the public rejection of GM food. Thirdly, at the same time there has been a growing movement towards greater involvement in governance in all spheres of public policy at both local and national levels. This is seen as a possible supplement to the inadequacies of representative democracy or the 'democratic deficit' [9]. Depending on who is doing it, involving the public can mean anything from an education programme, through consultation, participation in policy making to citizen's control [10].

Within this context public consultations have been used to serve different agendas; to demonstrate the need for more education, to feed into the process of policy formation, for example, to find out how receptive society is to a new application; to feed into the implementation process, for example to find out public concerns in order to allay them, or even as a legitimation strategy (to fulfil the obligation to consult). Consultations undertaken as academic research have provided a critique of how knowledge is produced, showing that lay people have different kinds of expertise which are contextualised and experiential. This is leading to a reconceptualisation of the relevant issues in evaluating technologies and science policy which is starting to be acknowledged in some parts of government [11]. Overall consultations can be typified according to the degree to which the aims and scope are predefined or left open and in the commitment or authority to feed the results into policy formation [12].

Media Coverage of the New Genetics

Discussions of media coverage are an integral part of ideas about the public understanding of genetics. The media is seen to be the prime source of information and as highly influential over public opinion, although this is difficult to demonstrate [13]. Public mistrust and suspicion of technology is often attributed to negative and hyped media coverage. Scientists and policy makers often complain about a lack of good science reporting but evidence suggests that there is no uniform pattern of 'science hostile' reporting [14]. A survey of coverage of genetic engineering in the US, UK, France and Germany media concluded that, on the contrary, the benefits were emphasised, there was little on disconfirming evidence and British coverage could 'almost be called euphoric' [15]. Science reporting in the popular press/magazines is 'a collection of success stories' with a lack of scepticism [16]. Uncertainties are not discussed and neither is the social context of the new genetics [17]. In the UK press the 'hard' science stories report breakthroughs in an uncritical way, often several times over the years.

Xenotransplantation Over the Years

'It is possible that selec-tion of patients would begin next year for the first transplants of the pig organs into humans'.
-Director of Transplant Services, Papworth Hospital, Cambridge, UK- co-founder of Biotech company Imutran quoted in The Times, 23 August 1995

'The moral question of using animals purely for organ donation had already been dealt with: 'I don't think a moral question can be tissue specific. You can't accept a pig's heart valve but not the heart'
-Director of Research, Imutran quoted in The Independent, 13 Sept 1995

'I think we are four or five year's away from meaningful human clinical trials'
-Head of Research at PPL Therapeutics quoted in The Guardian, 3 January 2002

This may account for the common complaint by experts that the public think genetic applications are already in use. Conrad and Weinberg chart a similar story in their article 'Has the gene for alcoholism been discovered three times since 1980? [18]. Journalists rely on established sources for these 'breakthrough' stories; that is, the mass-circulation science and medical journals, staged events organised by biotech companies and research institutions and comment from the researchers themselves and other scientists from prestigious institutions [19].

While genetic news is reported uncritically there is another side to the coverage once genetic technology is in use. Stories focussed on personal experiences, 'the human drama of genetics' then appear, particularly in women's magazines. Here the simplicity and certainty of hard science stories gives way to uncertainty and shades of grey [20]. These stories, for example, show that the discovery of a gene and possibility of genetic testing may raise difficult questions rather than provide solutions [21]. For many scientists this is not good quality reporting because it deals with opinion rather than useful information and reliable evidence [22]. In fact these 'soft' stories examine science and technology in practice rather than celebrating science and technology uncritically in 'an unrelenting discourse of progress' [23]. However, they tend to concentrate on specific cases rather than considering the broader social and political issues.

To summarise, despite the evidence that reporting on developments in the new genetics is unreflective, public consultations which allow open responses and are not centred on technical knowledge find lay people have a contextualised understanding of science and its applications [24]. However, in many public consultations (including consensus conferences and citizen's juries) there is a 'pre-framing' of the issues i.e. what counts as relevant in any given debate [25]. Organisers of consultations circumscribe the items to be discussed and issues of fundamental interest to the public may be glossed over e.g. discussing who should be allowed to access a technology but not whether it should be developed in the first place.

Gene Week

The Gene Week project was funded by the Wellcome Trust, a charity that funds biomedical research, mainly in the UK (distributing around 900 million euros a year, including part of the Human Genome Project). Recognising that the research they fund has far-reaching implications for society, in 1998 the Trust announced a new focus for its public understanding of science and education work. The stated aim is 'to stimulate and inform debate about the social and ethical implications of biomedical research...' with a commitment to feed the public's views into policy development [26]. So, like everyone else, they are recognising the imperative to involve the public.

The focus of Gene Week was to try out a novel method of consulting the public about genetics. During the week of 18th-22nd March 2002 a series of articles were published on medicine and genetics in the Lancashire Evening Post and on its associated website, with a total of around 5000 words. Readers were invited to send in their views, encouraged by a prize draw. The Lancashire Evening Post is the local daily paper in Preston, with a circulation of 57 000. The consultation was novel in several ways:

1. Gene Week aimed to target a local readership, in the hope of reaching groups of people not normally consulted. A large proportion of the readership is working class with only 8% from social classes A/B.
2. Although there is extensive coverage of genetic issues in the national press, this is not the case in local papers. In the year leading up to Gene Week there were a dozen articles mentioning genetic conditions or stem cells, all of which dealt with 'tragedies' in just 5 local families e.g. a man who died from Huntington's disease and a girl with leukaemia which mentioned that she might be helped by stem cells. There were no general pieces about genetic developments. This is relevant because about a third of LEP readers do not take a national daily paper.
3. In contrast to the usual reporting of these issues The Gene Week articles were written to highlight *current* research and uses, drawing on local materials; focus on society rather than the science; show the complexity of issues, avoiding simple dichotomies (hope/fear, miracle/tragedy); using different kinds of experts (from clinical geneticists and ethicists to a local person living with cystic fibrosis and deaf university students).
4. Media consultations usually invite audiences to vote yes or no in response to complex questions - two recent examples of web based consultations by the BBC (website linked to TV programmes) and Gene Forum. The Gene Week provided the opportunity for a large-scale consultation, whilst allowing individuals to respond as fully and widely as they wished.

Topics covered included genetic testing, the regional genetics service, pre-natal screening, pre-implantation genetic diagnosis, xenotransplantation, cloning, differing expert views, patients' experiences, biobanks, and commercialisation. The questions posed were very broad to allow people to tell us about anything that interested them.

This discussion will focus on the topics people chose to take up from the material presented and how they reasoned, in other words, how they did ethics. But, in general, responses ranged from one word ('no' to cloning) to lengthy discussions and readers picked up on every topic from the articles. They made links between topics in terms of the issues they wanted to raise and similar lines of arguments emerged from responses to different questions and articles. Because of this the responses were analysed in terms of the arguments rather than the topics. I am going to discuss three of the main themes with examples of responses; questioning the context of science, looking at how you frame autonomy in the messiness of reality and health care policy.

Theme One: Questioning the Context of Science

Respondent 016:
The people behind the research into genetics are just scratching the surface of what will affect all our lives, now, and for years to come. So they must tread carefully. Gene therapy trials carried out recently in the USA, (to fix a disease at its root in the DNA) have been bafflingly bad. Not one trial demonstrated that a disease could be cured or controlled by altering a person's genes. One of the main stumbling blocks for this lack of success is the fact that genes make proteins and proteins interact in ways we do not yet understand. Whether or not the public think genetic medicine is good or bad, moral or immoral, the research will continue regardless. Huge profits are hoped to be made by the pharmaceutical companies when cures are found for medical conditions such as cancer, heart disease, arthritis, diabetes and aids. All of which make attractive targets for these companies. We might be wise to think upon St Hildegard of Bingen's words of wisdom. "All of creation God gives to humankind to use. If this privilege is misused, God's justice permits creation to punish humanity."
e-mail response, male, 41-60, assembly worker

He is concerned about the power of science and the powerlessness of the public to influence research – research will continue regardless of anyone's ethical stance. The public doesn't get the chance to make decisions because research is driven by pharmaceutical companies attracted by profits. Although it's clear that he is concerned about the distant future and the need to tread carefully, he is not saying emphatically that he is for or against he technology but is questioning its governance.

Respondent 004: (anyone with teenage children will recognise the email style, stream of thought without much attention to spelling, capital letters and so on!), in response to a question on prenatal screening for genetic conditions: should you screen for any of these conditions?
who is to say whether it is right or wrong?? people may say its playing God, but then others may say it is what God made us to do, evolution and all that. however it may have considerable unforeseen consequence for future humans. it also has moralistic issues to think about as how could someone say that they want to aborut a birth as the kid maybe downsyndrom, when just because they may have this disability it does not mean they will have a horrible life, as downs as any other person finds joy in stuff. it is a very controversial topic and i dont think we will ever find the right and wrong of it, but we should weigh out the pros and cons to see if it is worth taking the risk. By allowing scientists to take control of the world by allowing them to have the ability to use genetics to

their advantage by giving them power and a lot of potential to make money!! its very dangerous !!
e-mail response, female, age 17, student

It is interesting that she starts off with are some recognisable arguments about autonomy and how to measure quality of life – you cannot just use standard measures of mental impairment because someone with down's 'finds joy in stuff'. Who is to say whether or not another's life is worth leading? But this leads her to think about 'the scientists' who are making decisions about prenatal screening. She is worried about scientists having the power, because the power itself and the potential for profit might lead to dire consequences ('is very dangerous').

Respondent 058:
To summarise my views, I think that some developments have frightening prospects and I'm glad that I won't be alive in the 22nd century! Activities must be strongly regulated, and I'm glad that I live in a society where 'what will be will be' still largely prevails. I can see some benefits of the genetic developments for some people where it will definitely increase their quality of life. But another of my main concerns is that genetic testing will become a big commercial business and that we will be influenced to behave in ways that major profit making companies tell us to.
postal response, female, 26-40, local government officer

More explicitly critical of genetic technologies- this is the end of an extract talking about her views on susceptibility testing for multifactorial conditions- she is glad that she won't be faced with the opportunity to use genetic technology, and wants to take life as it comes (what will be will be). Again, she is concerned about pharmaceutical companies driving the direction of research and influencing the way we behave – implying that if the test is there people might feel obliged to take it.
    Looking at these extracts, people do not see the point of thinking about what is right and wrong in the abstract without considering the social context in which people live and in which research is done, science produced and applied. They do not see the ethics of a particular case as being resolvable without looking at the broader context- ethics is about structures as much as individuals.

Theme Two: How Do You Frame Autonomy in the Messiness of Reality?

Respondent 014:
Morally I would say you shouldn't put any animal parts into humans but I think if it was a family member or close friend I would want them to have the best quality of life possible and if this meant living with a pigs heart or such like then I would have to agree.
e-mail response, female, 18-25, student

Respondent 062:
Similarly I would not want to choose the sex of my child even though I have three boys (sons) and badly wanted a girl. So perhaps it was a good thing that 40 years ago you couldn't tell which sex you were carrying!
postal response, female, 61+, retired

The extracts acknowledge that individuals might decide immoral things because they will benefit themselves or their families. The second makes the point more explicitly- we need to be protected from ourselves. People have moral conflicts because when something is available it is difficult to refuse if it is your family that is affected.

Respondent 051:
If we hadn't had our son before the genetic test was made I would have been recommended to have an abortion so at least the diagnosis made sense of his problems. But sometimes the wrong emphasis is put on statistics so you must be careful with results.
postal response, female, 41-60, carer and housewife

Someone with experience of a genetic condition unpacks autonomy – the stress in genetic testing is on personal choice (non-directive counselling, although she thinks she would have been 'recommended' to have an abortion). But how can you exercise autonomy and decide for yourself when you are dependent on the medical profession to provide the information (interpret the statistics)?

Theme Three: Health Care Policy

Again, these extracts show the public taking a broad perspective. They do not just talk about a decision being ethical or unethical, but highlight the need to look at the consequences of any decision in social context.

Respondent 007:
Genetic advances would allow us in the future to eradicate certain diseases etc. What is worrying me is who would regulate this practice and ensure it was not open to abuse. To eliminate some types of diseases would be obviously advantageous, but where would it end? Also would these "eradication" initiatives be backed by the government, in light of the fact that people would ultimately live longer etc. thus use more of the countries resources. Would the government not think this too expensive? There would be massive implications for the future economy, insurance industry and health care industry.
e-mail response, female, 26-40, ward manager psychiatry

We cannot just go on and eradicate diseases without thinking about the consequences in terms of resources – what happens if the country cannot afford it? Who will regulate the eradication and decide priorities?

Respondent 027 (responding to the story of Rachel, a 20 year old student with cystic fibrosis):
I'm not sure I'd take a test to find out my risk of anything because I think all it would do would cause fear. I'd rather think and hope that somebody like Rachel should have had their symptoms recognised earlier – no point in her case to think of having been tested for risk of cystic fibrosis earlier, because from the length of time it took to diagnose, it obviously hadn't occurred to anyone anyway! What I'm saying is I'd rather the NHS was on the ball enough to recognise symptoms as they occur...
postal response, female 26-40, artist

Thinking about testing for cystic fibrosis she brings up the ability of the National Health Service to recognise symptoms, and sees this as a prior question to whether testing should be carried out. The test was there but the health service failed Rachel (she was not tested until age 11).

Respondent 103:
With all the new technologies (such as this) we are allowing humans to live longer and longer. This will, inevitably, cause problems long term with over-crowding. If we carry on the way we are and keep everyone alive as long as possible, we may have to resort to some sort of culling or a limitation to life spans; this would not be gladly welcomed.
postal response, female, age 17, student

New technology will lead to new social and ethical problems- people tend to assume that longevity is desirable, if there is a good quality of life, but she speculates about possible social consequences (this tends to be something young people bring up!).

Respondent 017:
The collection and storage of stem cells from a newborn's umbilical cord makes sense, but it makes more sense if donated to a communal bank (rather like our existing blood banks) than if stored for personal use. The reasons are: 1. the individual for whom it is stored is unlikely ever to use it, so it goes to waste. 2. Quantities collectable from a single sample are small and unlikely to be effective beyond childhood. 3. once used, a sample collected for individual use is gone. 4. Individual collection and storage costs hundreds of pounds: hopefully, communal banking will be free. As techniques for matching and using UCB which is less exactly matched with the recipient continue to improve, so communal banking makes more sense. It has the potential to save many lives.
e-mail response, male, 41-60, NHS manager

This was a response to a paragraph on stem cell research from an NHS manager (the public have all sorts of expertise). Instead of discussing the ethics of commercial cord banks, which were mentioned in the article, he is again widening the perspective and posing an alternative solution of communal banking. This 'makes more sense' – using a utilitarian perspective- because while most single samples stored for individuals will never be used, a communal bank will benefit more people and be open to everyone instead of restricted to those who can pay.

To Sum Up This Analysis

People saw ethical issues in how technologies come to be developed and implemented in the first place and recognised that once a technology is in place it will be widely used in unanticipated ways. It is therefore important for society to be given the opportunity to consider the broader implications of technologies before they are funded and developed rather than making decisions only on a case by case basis. This research demonstrated again the value of open-ended consultations in which lay people have opportunities to define for themselves the relevant issues.

Why Should Research like this be of Interest to Ethicists?

The research points to a need to broaden the perspective to a wider range of concerns rather than being limited to the concepts deemed to be in the realm of medical ethics (traditionally individualistic concepts like autonomy, consent, and confidentiality) [27]. Bioethicists can learn from lay people what counts as relevant in decision-making. Given the new emphasis on public input into policymaking and a lessening of the emphasis on experts' views, there are implications for the role of ethicists as experts. Bioethicists already see their role in different ways, for example, as providing an answer to moral questions or as clarifying the issues and arguments.

So perhaps there is a role for bioethicists in clarifying and facilitating discussion of issues raised in public consultations.

It still remains to be discussed whether ethicists would like to fulfil this role as ethical facilitators and what ethicists believe to be the moral standing of decisions based on consultation/participatory democracy.

References

[1] Kerr et al, (1997), p.291
[2] Tony Blair, British PM, 20/5/02 Times p.1
[3] Voss, (2000)
[4] Wellcome Trust, (1998)
[5] Zimmerman et al, (2001), p.55
[6] Barns et al, (2000); Kerr et al, (1998)
[7] Joss, (1999)
[8] Dickson, (2000)
[9] Harrison and Mort, (1998); Cooper et al, (1995)
[10] Purdue, D, (1999)
[11] House of Lords, (2000), pp.16-24
[12] Harrison and Mort, (1998); Milewa, (1999); Petersen, (1999) p.27f
[13] Peters, (2000); Petersen, (2001), p.1267; Schenk and Sonje, (2000), p.331
[14] Kohring and Gorke, (2000)
[15] ibid, p.353
[16] Zimmerman et al, (2001) p.54
[17] Petersen, (2001) p.163
[18] Conrad and Weinberg
[19] Petersen (2001); Chadwick and Levitt, (1997)
[20] Henderson and Kitzinger, (1999)
[21] Chadwick and Levitt, (1997)
[22] Lewestein, (1995) p.344
[23] Henderson and Kitzinger, (1999) p.570
[24] Wynne, (1996); Kerr et al, (1998a), (1998b); Barns et al, (2000)
[25] For instance, Barns et al., (2000); Irwin, (2001)
[26] Wellcome Trust, (1998); (1999)
[27] Barnes, (2000)

Bibliography

Barns, I, Schibeci, R, Davison, A, and Shaw, R, (2000), "What Do You Think About Genetic Medicine?" Facilitating Sociable Discourse on Developments in the New Genetics, *Science, Technology and Human Values* 25 3 pp.283-308
Conrad, P, (1999), 'A Mirage of Genes' *Sociology of Health and Illness,* 21 2 pp.228-241
Chadwick, R, and Levitt, M, ' Mass Media and Public Discussion in Bioethics' in Chadwick, R, Levitt, M and Shickle, D, (1997) *The Right to Know and the Right Not to Know* Ashgate, Aldershot pp.79-86
Conrad P. and Weinberg D. 'Has the Gene for Alcoholism Been Discovered Three Times Since 1980? A News Media Analysis'. *Perspectives on Social Problems* 8 pp.3-24
BBC (2002) Hot Topics – Intelligence – Nature or Nurture? http://www.bbc.co.uk/science/hottopics/intelligence/clever.shtml 18/6/02
Cooper, L, Coote, A, Davies, A, and Jackson, C, (1995), *Voices Off: Tackling the Democratic Deficit in Health,* IPPR, London
Dickson, D, (2000), "Science and Its Public: the Need for a 'Third Way'" ,*Social Studies of Science* 30(6) 917-23

European Commission (1996) *Eurobarometer 46.1* Report
European Commission (1998) *Eurobarometer52.1* Report Question 4
Hampel, J, and Renn, O, (eds), German Attitudes to Genetic Engineering , Special Issue *New Genetics and Society* 19:3
Harrison, S, and Mort, M (1998), "Which Champions, Which People? Public and User Involvement in Health Care as a Technology of Legitimation." *Social Policy & Administration* 32(1), p.60-70
Henderson, L, and Kitzinger, J, (1999), 'The Human Drama of Genetics: 'Hard' and 'Soft' Media Representations of Inherited Breast Cancer' *Sociology of Health and Illness* 21 5 pp.560-578
House of Lords (2000) *Science and Society* Select Committee on Science and Technology HL Paper 38 The Stationery Office, London
Human Genetics Commission (2001) *Public Attitudes to Human Genetic Information*, People's panel quantitative study conducted for the Human Genetics Commission, London, HGC
Irwin, A, (2001), 'Constructing the Scientific Citizen: Science and Democracy in the Biosciences, *Public Understanding of Science* 10 pp.1-18
Joss, S, (1999), "Public Participation in Science and Technology Policy and Decision-Making - ephemeral Phenomenon or Lasting Change?" *Science and Public Policy* 26(5) pp.290-293
Kerr, A, Cunningham-Burley, S, and Amos, A, (1997), The New Genetics: Professionals' Discursive Boundaries, *Sociological Review* pp.278-303
Kerr, A, Cunningham-Burley, S, and Amos, A, (1998a), Drawing the Line: an Analysis of Lay People's Discussions about the New Genetics, *Public Understanding of science* 7 pp.113-133
Kerr, A, Cunningham-Burley, and Amos, A, (1998b), The New Genetics and Health: Mobilizing Lay Expertise, *Public Understanding of Science* 7 pp.41-60
Kohring, M, and Gorke, A, (2000), Genetic Engineering in the International Media, *New Genetics and Society* 19 3 pp.345-363
Levitt, M, Public consultation in bioethics. What's the Point of Asking the Public When They Have Neither Scientific Nor Ethical Expertise? Submitted to *Healthcare Analysis*
Lewestein,B, (1995), Science and the Media, S, Jasanoff, G, Markle, J, Petersen, T, Pinch (eds) *Handbook of Science and Technology Studies*, Sage Publications, California, London. Revised edition pp.343-360
Milewa, T, (1999), *Public Opinion and Regulation of the New Human Genetics: a CriticalRreview*, Canterbury University of Kent, Centre for Health Services Studies
People Science and Policy Ltd, (2002), *BioBank UK: A Question of Trust,* People Science and Policy Ltd London. Report prepared for the Medical Research Council and The Wellcome Trust.
Peters, H.P, (2000), The Committed Are Hard to Persuade. Recipients' Thoughts During Exposure to Newspaper and TV Stories on Genetic Engineering and Their Effect on Attitudes, *New Genetics and Society* 19 3 pp.365-381
Petersen, A, (2001), Biofantasies: Genetics and Medicine in the Print News Media *Social science and Medicine* 52 1255-1268
Pfister, H-R, Böhm, G and Jungermann, H, (2000), The Cognitive Representation of Genetic Engineering: Knowledge and Evaluations, *New Genetics and Society* 19 3 pp.295-316
Purdue, D, (1999), "Experiments in the Governance of Biotechnology: a Case Study of the UK National Consensus Conference" *New Genetics and Society* 18 1 pp.79-99
Schenk, M, and Sonje, D, (2000), Journalists and Genetic Engineering' *New Genetics and Society* 19 3 pp.331-343
Voss, G, (2000), *Report to the Human Genetics Commission on Public Attitudes to the Uses of Human Genetic Information,* London Human Genetics Commission
Wellcome Trust, (1998), *Public Perspectives on Human Cloning*, London Wellcome Trust
Wellcome Trust, (1998), Right and Responsibilities: Engaging the Public in Informed Debate, http://www.wellcome.ac.uk/en/1/mismiscnernr.html: 13.08.02
Wellcome Trust, (1999), The Medicine in Society Programme http://www.wellcome.ac.uk/en/1/mismis.html 16.11.01
Wynne, B, (1996), Misunderstood Misunderstandings: Social Identities and Public Uptake of Science, A, Irwin and B, Wynne (eds) *Misunderstanding Science?* Cambridge, Cambridge University Press pp.19-46
Zimmerman, C, Bisanz, G, Bisanz, J, Klein, J, and Klein, P, (2001), Science at the Supermarket: a Comparison of What Appears in the Popular Press, Experts' Advice to Readers, and What Students Want to Know, *Public Understanding of Science* 10 pp.37-58

The Use of Computer Simulation and Artificial Intelligence in the Study of Ethical Components of Medical Decision-Making

Peter ØHRSTRØM,
*Department of Communication,
Aalborg University*

Jørgen ALBRETSEN,
*Humanistic Information Science,
University of Southern Denmark*

Søren HOLM,
*Institute of Medicine, Law and Bioethics,
University of Manchester*

In this paper we intend to study how computer simulation and methods from artificial intelligence may be used in the study of medical decisions in real life. In section 1 we consider a number of methodological problems in the study of ethical decision-making. We consider problems in relation to more classical methods as well as the methodological problems to which the use of computer simulation may give rise. In section 2 a particular computer simulation, KARDIO, is introduced. This simulation has been developed in order to study medical decisions, and it is argued that the pilot studies which have been carried out so far may be taken as a clear indication of the usability of computer simulations in the empirical study of medical decisions. In section 3 we discuss the possibility of a new KARDIO module using techniques from artificial intelligence.

1. Methodological Problems in the Study of Ethical Decision-Making

As Søren Holm and others have argued [1], the study of medical decisions in real life obviously gives rise to a number of methodological problems of both a practical and research ethical nature. There are often practical problems in being allowed access to the medical settings, and data collection may also be complicated. To limit the time needed for data collection, many studies are done on the basis of medical records. This raises methodological problems, since the information in medical records is an edited/reconstructed version of what really happened. If the researcher is present in the medical setting it does, however, also create methodological problems because the mere fact of his or her presence may change the decision-making process in unpredictable ways. The research ethical problems are connected to the need to obtain informed consent from doctors and patients. Even though these studies are primarily interested in the thought processes and decisions made by the doctors, the patients are inevitably drawn in, since what makes these decisions interesting is mainly that they are made about patients and based on information about patients. It is impossible to study the decision-making of doctors without making the patients subsidiary study objects.

These problems have led a number of researchers in the field of empirical ethics to pursue a methodology utilising case-based questionnaires, i.e. questionnaires where a short clinical case is presented, and where the respondents are asked to answer a number of questions about what they would do in such a situation.

Some studies attempt to build in a temporal progression in the cases. This creates technical difficulties in a paper-and-pen questionnaire, because it is difficult to prevent the respondent from looking ahead to see what comes next in the case. It is also difficult to provide branching opportunities depending on the choices the respondent makes.

The case methodology is usually chosen in order to try to get behind the values doctors claim to have to the values they really have, and which are demonstrated by their responses to cases. It would be possible to try to ask doctors directly what their values are, but such a study would encounter serious problems with respondents knowing the socially acceptable answer. Very few physicians would ever admit that socially unacceptable values influenced their decisions (e.g. racist or ageist values), even if they did.

A computer simulation seems the obvious solution to the problems of case-based questionnaires in the study of the ethical aspects of medical decision-making. It does not create the same practical and ethical problems as a study of a real life setting, and it can, if the simulation is suitably constructed, present the respondent with a "naturally" developing scenario, where there is no possibility of looking ahead, whilst at the same time being responsive to the decisions the respondent makes.

The use of computer simulations does, however, give rise to a number of other problems. The features of an ideal simulation cannot be described in general, but it will depend on the purpose for which the simulation is being designed [2]. If the intended use of the simulation is to train a person to perform a certain complex set of cognitive and motor skills (e.g. a cockpit simulation intended for pilots) a certain type of very life-like simulation is needed (real-time, full scale etc.). If the purpose of the simulation is more modest it may be neither necessary nor prudent to aim for absolute similarity to real life.

In the following we want to study the problems related to systems like KARDIO, which is the name of a computer simulation system we are developing. KARDIO is a computer simulation of a cardiac care unit. More details of the system will be presented in the next section.

With respect to systems like KARDIO, we may at least consider two kinds of problems related to the idea of simulation. Firstly, there are the methodological problems of translating real life situations into situations in the computer scenario and vice versa. Secondly, there are the problems of the validity in real life of results obtained by studying the decision-making in the computer scenario. These problems are illustrated in figure 1.

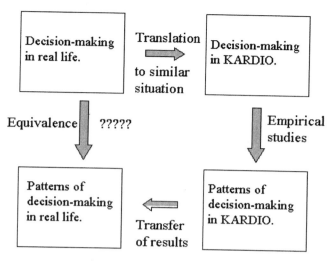

Fig. 1. The simulation problem.

The translation problem may also be seen as a representation problem. How can a real life situation be represented in a formal way? In order to build a simulation system representing the real life setting it will be necessary to have a description of the virtual scenario formulated in some appropriate formal or semi-formal language. Moreover, a notion of causality is needed. The implementation of this notion of causality must include information about which events necessarily follow from and after any possible situation in the virtual scenario of the system. Furthermore, the system must include information regarding the alternative possibilities for each possible situation, which are left open for the user's choice. We suggest that all this can be obtained by using temporal logic [3]. This means that "naturally" developing computer scenarios may be represented and studied in terms of concepts and ideas from temporal logic. In addition, we suggest that results in the computer scenario stated in terms of temporal logic are in fact translatable into natural language in a straightforward manner.

How can we know that the results of the empirical studies of the use of the simulation are relevant when it comes to the real life situation? Although we cannot give a definitive proof of the equivalence of the results of the simulated setting and the real life situation, we can in fact present some arguments and some relevant tests. First of all it should be mentioned that interviews made in our pilot studies show that all relevant real life procedures have their parallel in the present version of the simulation system. The users of this version of the simulation system in general simply find that all the actions and procedures which they would like to carry out, can in fact be carried out in the simulation system. This is of course due to the fact that KARDIO has been developed iteratively involving the users in all phases of the development process. During this process we have implemented all the realistic improvements of the system suggested by the users based on their testing of the various versions of the system. The interviews of the users of the pilot studies also show the user-interface is now fully accepted by physicians. According to the users' statements, the simulation system is sufficiently life-like to be valuable in the study of decision-making. Observation of the users also show that in general the users are willing to run the simulation for up to 2 hours, and they normally become very involved in treating their "patients". The users obviously see the system as a realistic platform for testing their professional abilities as physicians.

2. KARDIO: A Simulation of a Cardiac Care Unit (CCU)

KARDIO is a computer simulation of a 6-bed cardiac care unit. An earlier version of the system has been described by Holm [4]. This simulation system is based on models of a limited number of possible cardiac diseases. In this way a virtual universe has been established. It is 'inhabited' with virtual patients, and it is equipped with a number of 'laws of nature' as well as with a certain degree of randomness. This means that the user can play the role of a doctor, in the sense that he can treat the 'patients' according to his decisions. In KARDIO the reactions of a patient are, on the whole, predictable over time, given the disease and the treatment. In this system the user is actually supposed to be a physician from real life and it is assumed that he will make decisions in the virtual scenario which are similar to the decisions he would make when he had to treat real patients in a real cardiac care unit.

The KARDIO interface (see figure 2) is made as graphic as possible, although there is at present no real-time video or speech involved. Under each virtual bed there is a field displaying the name and age. The "beds" are clickable. In this way the user has an easy access to the patient records. Other basic information such as pulse, blood pressure, and pain level of the patient in the bed can also be found here. The interface also contains a clock displaying the remaining time for the present ward round and buttons for ending a ward round, or ending the program.

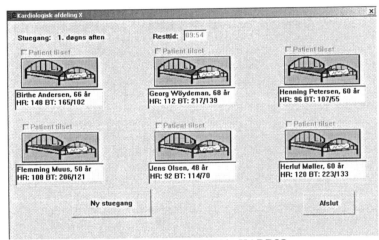

Fig. 2. The virtual CCU in KARDIO

The simulation can display a number of data about the individual (virtual) patient (see figure 3). A picture of the patient is presented together with a field showing the patient's name, occupation, and age. There is also a scrollable patient record and a continuous one-lead monitoring ECG. The patient record window is designed to give a quick overview of the state of the patient, and enable the physician to make initial decisions about further information gathering or initial treatment decisions.

The system has been developed as a possible tool for the study of ethical and other aspects of medical decision-making. The system is designed to be quite realistic in order to make it likely that the reaction pattern of the physician who is running the system will be similar to the pattern in real life. If such a similarity can be assumed, then the simulation system provides a nice tool for the study of decisions concerning individual patients, as well as decisions concerning prioritisation.

In order to obtain data relevant to the analysis of ethical values, certain decisions have been made obligatory in the simulation. The construction of KARDIO is partly based on information gathering from real life. This work has, for instance, shown that it is routine in many CCUs in Denmark to decide whether or not a given patient should be resuscitated, and whether or not the patient can be moved to an 'ordinary' medical ward, following each ward round. Therefore these two decisions have been made obligatory, since they provide a pointer to the priorities of the physician running the simulation.

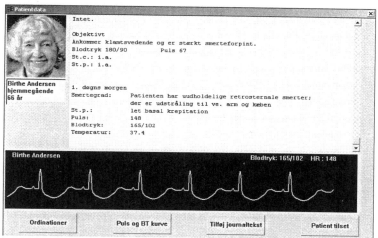

Fig. 3. The patient record in KARDIO

It is very clear that this simulation system is based on a rather complicated structure of events involving a mathematical model of the development of various diseases over time and the reaction to various possible treatments. The system also gives rise to a description in terms of temporal logic of the many decisions and choices the user has to carry out when running the system. In this way the model corresponds to a branching time system.

As mentioned above, the need for description using temporal logic makes it necessary to see the virtual scenario of a simulation system like KARDIO as an extension of the real world rather than as a separate world. The understanding of the virtual scenario as an extension of the real world rather than as a new world becomes even more evident when we are discussing systems (like KARDIO) within which ethical choice is simulated. In this case the virtual scenario not only shares its 'nowness' with the real world. Some or perhaps even most of the values of ethical importance are also supposed to be shared with the real world.

The possibility of allowing the KARDIO user to 'undo' his decisions when he realises that a certain treatment leads to unwanted results may be considered. Such a feature would of course make the system a bit unrealistic. It is, however, an open question as to what extent such a facility would reduce the possibilities of transferring the results regarding ethical behaviour from the virtual scenario presented by KARDIO to the scenario in a real CCU at a real hospital.

3. The Development of a New KARDIO Module using Artificial Intelligence

As we have seen above, KARDIO is designed to simulate not only some medical reactions and processes but also the ethical perspectives related to this medical reality. It is assumed that the use of the system can, to a large extent, generate the same responsibility and ethical judgements as the reality which is being simulated. It is our hope that the users will experience the same kinds of ethical dilemmas as in real life.

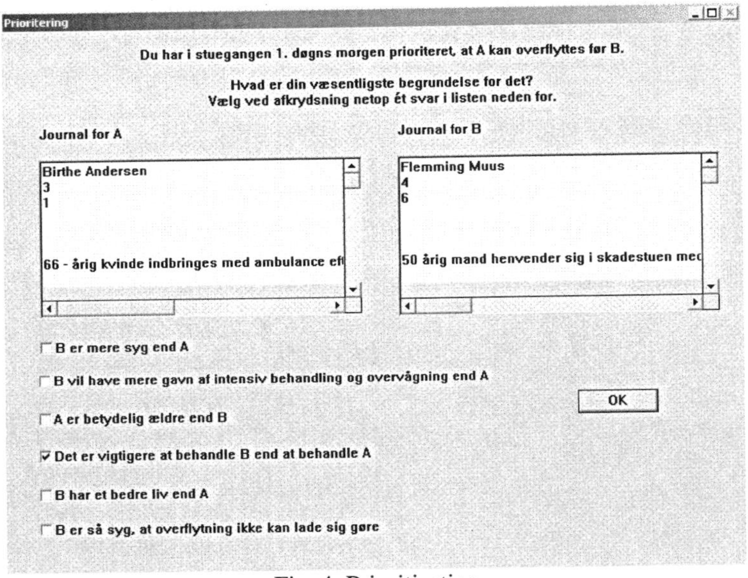

Fig. 4. Prioritisation

It is well known that many ethical dilemmas involved in medical decision-making relate to prioritisation. An attempt has been made to implement such decisions in KARDIO. When running the simulation system the user is asked to carry out a prioritisation of the six patients in the unit. After a session the system will ask a number of questions concerning this prioritisation. This is done in the scheme illustrated in *Fig. 4* above, in which the user has to state why he or she has chosen a certain prioritisation order.

In *Fig. 4*, the user is asked to explain why he or she has stated that patient A should be transferred to his or her home or to less intensive care instead of patient B. It is assumed that the user will be able to point to one (and only one) of the following claims as his or her main reason:
-B is sicker than A
-B would benefit more from intensive care and treatment than A
-A is considerably older than B
-It is more important to cure B than to cure A
-B has a better life than A
-B is so sick that transfer is not possible

    As a working hypothesis, it is assumed that every physician will carry out his prioritisation based on a fixed order of the importance of the elements in the above list of claims. This order is supposed to reflect the physician's professional values. This way of seeing things represents an assumption with some empirical bearings, and it should obviously be tested empirically. It will be interesting to know to what degree the users are able to stick consistently to a fixed order of the importance of the above claims. In the following, however, we will take the assumption for granted. If later empirical studies show that the assumption is too far from reality, the logical model has to be adjusted accordingly.

We may model various types of professional values based on which the users may act. As an illustration of the procedure we'll present three virtual agents with whom the users may be compared:

1) *The deontological doctor (d1)* thinks in terms of classical medical virtues. He thinks the physician should treat his patient in a way that is purely based on medical considerations. Who the patient is and what his quality of life might be when he (hopefully) returns home should not, in general, be taken into account. The age of the patient in itself should not normally be a determining factor for the physician.

2) *The empathic doctor (d2)* thinks that patients should be prioritised according to how important it is for them to be treated. He thinks that he may be able to estimate this degree of importance when trying to identify with the total life situation of the individual patient (including the patient's background in social life). This virtual doctor tries to answer the question: How should I be treated, if I were in the position of patient A in comparison with how I should be treated if I were in the position of patient B?

3) *The utilitarian doctor (d3)* thinks that the restriction of resources must be taken into very serious account. The medical resources should, in his opinion, be used in order to create as much quality of life as possible. For this reason it is more important in most cases to save the life of a young person that it is to save the life of an old person. It should also be taken into account that some patients are more important to other people's quality of life than others are.

In KARDIO the 3 virtual doctors corresponds to the following lists of the claims:

The deontological doctor (d1)	The empathic doctor (d2)	The utilitarian doctor (d3)
B is so sick that transfer is not possible	B is so sick that transfer is not possible	B is so sick that transfer is not possible
B is sicker than A	It is more important to cure B than to cure A	B would benefit more from intensive care and treatment than A
B would benefit more from intensive care and treatment than A	B has a better life than A	A is considerably older than B
It is more important to cure B than to cure A	B would benefit more from intensive care and treatment than A	It is more important to cure B than to cure A
B has a better life than A	B is sicker than A	B has a better life than A
A is considerably older than B	A is considerably older than B	B is sicker than A

For each of the virtual doctors (d1, d2, d3) the "list of value claims" shown indicates the order in which the various prioritisation criteria should be used in the KARDIO system.

The above lists of claims should be integrated in a logical framework dealing with the logic of obligation for use in KARDIO. However, it has turned out that the establishment of such a framework is a very difficult task. There is obviously a lot to do in order to establish a fully-fledged theory in terms of which of the relevant features of the virtual scenario can be described. For the present we have to make do with a rather limited theory.

Since the 1940s several logicians have tried to find a satisfactory logic of obligation formulated in terms of modern symbolic logic (i.e. a deontic logic), and there are still many unsolved problems in this field. One of the first logicians who dealt with the problem was A.N Prior (1914-69). In his book *Logic and the Basis of Ethics* Prior wrote:

... it remains true that a mistake about either (a) the kind of situation we are in, or (b) the obligation to which that kind of situation gives rise, can and often does lead, by a valid process of inference, to a mistaken view of our present categorical obligation; and also that our true present obligation could be automatically inferred from (a) and (b) if complete knowledge of these were ever attainable. [5]

Here Prior clearly states the main problems with which we have to deal if we want to implement a system from which our present obligation can be automatically inferred. We would need access to a complete knowledge not only of all valid ethical principles but also of all relevant data about the present situation. In everyday life such a complete knowledge would not be very likely to be attainable. For this reason we often have to make do with incomplete knowledge or with beliefs. This only makes things more complicated.

It is obvious that the ethical character of an act or a series of acts performed by a physician using KARDIO should not only be evaluated on the basis of a set of objective or general ethical norms. It will also be relevant (and sometimes even necessary) to take the user's knowledge and beliefs into account. For such reasons, various ideas of modality are needed in order to give a formal or semi-formal description of simulation systems like KARDIO. This topic has been thoroughly investigated by Marie-Laure Ryan [6], who has published an interesting analysis of the modal structure of the narrative universe as such. According to Ryan a narrative system outlines an actual world and a number of alternative possible worlds. In fact, Ryan has suggested that (in addition to plain "possibility") at least four kinds of modality should be taken into account in order to establish a general and satisfactory semantical model, which can capture the modal complexity of narrative and interactive systems:
-K (representing belief and knowledge)
-O (representing moral obligation and permission)
-W (representing wishes and desires)
-I (representing goals and plans)

All of these kinds of possible worlds could be relevant within the context of a system such as KARDIO. For our present purpose, however, only a part of this vocabulary is needed. We shall limit the discussion to the use of the following modal operators:
B (read: "the agent believes that ...")
M (read: "it is possible that ...")
O (read: "it ought to be that ...")

We also need a predicate 'act' which takes as its arguments the set of instants and the set of possible acts that can be carried out in the present situation. The proposition $act(i,i',a)$ is read "at the instant i the instant i' can necessarily be obtained by performing the act". It should be noted that the instant i' is supposed to be one time unit later than i. It should also be noted that we are referring to A.N. Prior's idea of instants, according to which an instant is a complete description of the world situation including the past and the possible futures [7].

Things become more complicated if several agents are involved. In this case we have to introduce references to the identities of the various agents who are active in the virtual scenario. For instance, we may use the expression *perform(x,a)* as meaning "the agent x performs the act a".

In most deontic logics it is a valid thesis that if a certain act ought to be performed then it is possible to perform this act. This is sometimes called Kant's principle [8].

$O(perform(d1,a)) \supset M(perform(d1,a))$ (and similarly for $d2$ and $d3$)
By transpositions it follows that
$\sim M(perform(d1,a)) \supset \sim O(perform(d1,a))$ (and similarly for $d2$ and $d3$)

This may in fact be given as our reason for putting "B is so sick that transfer is not possible" on top of the list of value claims for all three virtual doctors.

In the case of many users acting in the virtual scenario, we also have to relate knowledge, beliefs and wishes to the identities of the agents (users). In our context the crucial operator will be:

$B(x,q)$ (read: "the agent x believes that q")

In order to implement the virtual doctors mentioned above, we have to define functions, $d1$, $d2$, and $d3$, from the set of Priorean instants into the set of possible actions. This means, that for an arbitrary Priorean instant i, $d1(i)$, $d2(i)$ and $d3(i)$ will be the well-defined actions which the three virtual doctors may choose to carry out at the instant (i.e. the situation), i, in which they believe themselves to be. It seems that the relation between such functions and the tense-logical set-up can be formulated in the following way:

$(i \wedge a=d1(i)) \supset O(perform(d1,a))$ (and similarly for $d2$ and $d3$)

When this is implemented, the actual user may be compared with $d1$, $d2$, and $d3$. A match between the user u and the virtual doctor $d1$ is found in case the following holds for any i:

$(B(u,i) \wedge a=d1(i)) \supset perform(u,a)$
(and similarly for $d2$ and $d3$)

It is a complicating aspect of this model that the user normally does not know exactly which instant is the present one since the complete state of affairs is normally unknown for the agent. Instead, we have to deal with some kind of ethical reasoning based on incomplete knowledge and beliefs. As first option in the program we may, however, assume that the physicians know (correctly believe) what situation (i) they are in. This means that $B(u,i)$ holds if and only if i holds. In the actual implementation this equivalence will only be questioned if it leads to absurdities. In such cases the program should look for solutions assuming that the physician is mistaken about the actual situation.

An obvious ambition for the further development of KARDIO would be to implement a larger number of virtual agents with which the actual user can be compared. If such a virtual inhabited world can be successfully implemented as an extension of the real world, it will be possible to evaluate the ethical behaviour of the physician who is using the system automatically, comparing his choices with the various ethical positions which have been implemented. However, a number of problems have to be solved before such a goal can be reached. In particular, the choice of the virtual doctors that should be implemented in the system should be based on empirical studies of the present version of KARDIO.

4. Conclusion

It is evident that interactive systems like KARDIO should be described using temporal logic. If we want to take the choice of the user into serious consideration, a logic of earlier and later will be insufficient. We have also argued that interactive systems share their 'nowness' with the real world, for which reason the virtual scenarios of interactive systems ought to be seen as extensions of the real world and not as independent worlds. Moreover, we have argued that this conclusion becomes even more evident if we are dealing with simulation systems such as KARDIO in which the focus is on ethical behaviour and ethical values. Finally, we have also seen that in order to describe the kind of features which could be imagined for the further development of systems like KARDIO, a number of modal operators will be needed in addition to the basic temporal logic. Obviously, there is still a lot to do in order to establish such a tense-logical description language. In particular, the implementation of a satisfactory model for the user's belief and knowledge may turn out to be extremely difficult. It seems rather likely, however, that a lot can be gained from empirical studies using the present KARDIO version as a tool. Such empirical studies may even lead to the development of an improved version of the simulation model and of the use of techniques from artificial intelligence studies.

Endnotes

[1] Holm, S et al (1993); (1996); (1997), (1999)
[2] See Øhrstrøm, P et al. (1991); (1992), (1996)
[3] Øhrstrøm, P and Hasle, P (1995) *Temporal Logic. From Ancient Ideas to Artificial Intelligence*, The Netherlands: Kluwer Academic Publishers.
[4] Holm,S et al.(1999), (2000)
[5] Prior, A.N (1949), *Logic and the Basis of Ethics*, Oxford, Oxford University Press p.41-2
[6] Ryan, Marie-Laure (1991) *Possible Worlds, Artificial Intelligence, and Narrative Theory*, USA, Indiana University Press, p.111 ff
[7] For further details on the Priorian concept of instants we refer to Øhrstrøm, P and Hasle,P (1995), op cit, pp. 221 ff
[8] see Prior A.N (1962), op cit, p.224

References

Holm, S, (1995), The Medical Hierarchy and Perceived Influence on Technical and Ethical Decisions, *Journal of Internal Medicine*; 237, pp.487-92
Holm, S; Gjersøe, P; Grode, G; Hartling, O; Ibsen, KE; Marcussen, H, (1993), Ethical Reasoning in Mixed Nurse-Physician Groups, *Journal of Medical Ethics*, 22, pp.168-73
Holm, S, (1997), Metodi Standardizzati per la Rilevazione dei Dati in Bioetica. In: Gindro, S, Bracalenti, R, Mordini, E, (Eds.). *La Ricerca in Bioetica: Politiche, Metodi, Strategie*. Rome, CIC International, pp.41-46
Holm, S, Øhrstrøm, P, Rossel, P, Pedersen, SA, (2000), Cognitive Studies of Ethical Reasoning Based on the KARDIO-simulator, *Mathematical Modelling in Medicine*, The Netherlands: IOS Press, pp.217-227
Holm, S, Øhrstrøm, P, Donner, C, (1999), KARDIO - A Simulation of a Cardiac Care Unit Intended for the Study of the Ethical Components of Medical Decision-Making, *Proceedings of the 12th International Florida AU Research Society Conference*, p.18 - 23
Prior, AN, (1949), *Logic and the Basis of Ethics*, Oxford: Oxford University Press
Prior, AN, (1962), *Formal Logic*, Oxford: Oxford University Press
Ryan, Marie-Laure, (1991), *Possible Worlds, Artificial Intelligence, and Narrative Theory*, USA: Indiana University Press
Ryan, Marie-Laure, (1998), *Cyberage Narratology. Computers, Metaphor and Narrative*, p.29 (unpublished)
Nielsen, Finn & Øhrstrøm, Peter, (1991), *Design and Implementation of a Symptom Language used for Diagnosing Component Faults in a Nuclear Power Plant*, RISØ-I-582, Risø National Laboratory, p.10
Øhrstrøm, P, Hasle, P, (1995), *Temporal Logic. From Ancient Ideas to Artificial Intelligence*, The Netherlands: Kluwer Academic Publishers
Øhrstrøm, Peter, Nielsen Finn R, Pedersen Stig Andur, (1992), Diagnosing Component Faults in a Generic Nuclear Power Plant Using Counterfactual and Temporal reasoning. In *Proc. from 4th MOHAWC Workshop*, RISØ 1992, p.62-87
Øhrstrøm, P, & Andersen Henning Boje (1996), *SCENARIO ASSISTANT - Using a Tense Logical Language to Handle Scenarios Supporting Training of Coordination and Communication* RISØ-R-918 (EN), ISBN: 87-550-2214-6. ISSN: 0106-2840. p.13

Conclusion

Empirical Research in Bioethics: Report for the European Commission

Søren HOLM and Louise IRVING
Centre for Social Ethics and Policy
University of Manchester

1. Introduction

Bioethics involves the deployment of principles and arguments in the resolution of ethical dilemmas arising from the life sciences, including medicine. Most ethical principles rely on, or assume, empirical premises (i.e. premises about 'the state of the world'). Most ethical arguments in bioethics also contain some empirical premises. These premises may be of several kinds. Some are more or less straightforwardly derivable from the relevant scientific literature, but others are mainly of interest because they illuminate ethically important questions, and may not be as readily available. These latter kinds of facts are often of a sociological or legal nature.

In the process of developing methods for gaining relevant empirical information in bioethics, a number of conceptual and methodological problems have become evident.

The main conceptual problem concerns the relation between empirical studies and normative argument in bioethics. It is traditionally held in moral philosophy that it is impossible to derive an ought from an is, that how the world is tells us nothing about how it should be. The idea of empirical studies playing a major role (or any role) in bioethics is therefore seen as problematic by some bioethicists.

The main methodological problems are centred on clearly defining the type of relevant empirical information, the type of relevant study population, and the type of methodology suitable for producing this information. These methodological problems are exacerbated because the field of empirical studies in bioethics attracts both bioethicists who often have limited knowledge about research methodology, and social scientists who often have limited knowledge about ethics. Many studies in the field are therefore either methodologically or philosophically naïve. This obviously has implications for the validity of research results, and for the use which can be made of them.

The EMPIRE project was conceived with the objective of providing a comprehensive account of these problems, and an in-depth analysis of the possible solutions. This report offers a summary of the concerns related to the use of empirical research to inform ethical argument, and suggests ways in which these concerns may be overcome.

2. The Problem of the Use of Empirical Research in Bioethics

There has been a lengthy, ongoing debate about the methodology and goals of bioethics. The Empire project sought to clarify and inform one aspect of this debate – the use of empirical data and research - and in the process, perhaps, to illuminate a better way of engaging in the bioethics endeavour.

There are a number of issues, assumptions and misunderstandings that need to be addressed before a discussion of the use of empirical research in ethics can proceed [1]. The main complication is that philosophical bioethicists tend to be very sceptical about the use of empirical data in the ethical reasoning process, for a number of reasons. Firstly, 'facts' about the world or existing states of affairs give no guidance on what *should* be the case. In other words, the world as it is tells us nothing of the world as it should be. Furthermore, philosophers believe that bioethics is about giving good moral answers to moral dilemmas; they do not necessarily see why ethicists need to understand the context in which their advice is given. Finally, philosophers generally consider that there is little to gain in canvassing the ethical opinions of the general public, so to engage in quantitative or qualitative research to canvass opinion is not considered the same worthwhile task as it perhaps is in social science.

It should be noted that the concerns run the other way also. Social scientists feel that moral philosophers can only be of limited use in the bioethics enterprise as problems in the field are situated and, by their nature, affected by multiple aspects of 'the world as it is' and are therefore too complex to merely apply abstract principles or frameworks to.

Philosophers who are removed from social science research often misunderstand its nature. Social science (and science, for that matter), is not a matter of accumulating 'facts', or of facts being 'out there' waiting to be discovered, but is necessarily constructed of theories and explanations of why people act as they do. .

Resolution of these problems requires explicit definition of empirical research. The working definition of the terminology of empirical research advanced by Bennett and Cribb, which was utilised by the EMPIRE Project, is outlined below.

3. How should empirical research be understood?

Because the disputes between moral philosophers and social scientists about the use of empirical research have been framed as a dichotomy, empirical research has generally been portrayed as a 'substitute' for moral philosophy, or as a servant of it. Bennett and Cribb [2] suggest that neither conception is helpful, and that clarity can only be achieved through an assessment of the nature of empirical research. Empirical research, for these purposes, is specified to be research conducted through all the social science disciplines. Bennett and Cribb identified three different pictures of empirical research.

1. All social science – i.e. all of the social science disciplines, all of the forms of enquiry they deploy, and all of their products. This covers a vast range of epistemological styles and approaches, including what would typically be described as 'theoretical' as well as 'empirical' work; and work which springs from very different concerns and interests – explanation, prediction and control, understanding, critique etc.

2. A circumscribed set of approaches to qualitative and quantitative data collection and analysis – i.e. the kinds of 'methods' that are discussed in social science research methods texts (e.g. surveys, quasi-experiments, case-study, ethnography etc.) There is no definitive way of drawing this boundary; it could, for example, reasonably include historical methods. What these approaches would share is that they involve systematic attempts to sample a 'piece' or 'slice' of the social world and 'find out about' and 'make sense of it'.

3. Any research that includes an empirical component (as defined in 2) – this would be one way of compromising between 1 and 2. It would exclude social science work

which was essentially theoretical or argumentative in nature, and which did not include any systematic use of empirical data and data analysis.

This demonstrates the myriad of ways in which empirical research is conceptualised. Some philosophers use the term to refer to any work which makes claims about social reality (as opposed to philosophical work). Social scientists tend to use it more narrowly to distinguish their empirical methods – to some extent at least – from their more theoretical methods.

Bennett and Cribb conclude that empirical methods are best understood as a 'surface' feature of the traditions of the particular disciplinary inquiry. Data collection and analysis is not the accumulation of 'theory free facts' but depend on theoretical frameworks and models. Thus, bioethicists' use of empirical research simply to provide empirical findings is analogous to policy-makers looking to moral philosophers simply for their ethical conclusions; as if (a) these conclusions carried weight independently of the quality of the theorising and argumentation that underpinned them and (b) this, theorising, argumentation and the other properties of philosophical discourse was not part of what moral philosophers have to offer to others. So, in the same way, the findings from empirical research need to be understood and evaluated in their contexts. It may be that these methodological and disciplinary excursions are more valuable to moral philosophers than the actual findings. For example, a study which looks at 'communication, class and consent' in doctor/patient encounters may have value primarily because it illuminates issues of class and social hierarchy in health care rather than because it yields robust 'hard data'.

The conclusion that Bennett and Cribb reach about empirical research is that it should be seen as analogous to moral philosophy. Whereas moral philosophy is the disciplinary project of making defensible ethical claims, the disciplines of empirical research combine into the project of making defensible claims about (describing, understanding, explaining) the social world and social life. This account of empirical research suggests that it is neither a substitute for moral philosophy nor does it merely service it – it has its own distinctive orientation and purpose.

Scepticism about Empirical Research

Of course it cannot be that all empirical research is accurate, valid, ethical or even truthful. There are many possible reasons for scepticism about empirical research, some trivial and some substantive. It is often done badly. There are many methods, all with their own particular flaws, to choose between. But the same criticisms could be made of moral philosophy. The less trivial reasons are the inherent difficulties involved in making knowledge claims about the social world. Bennett and Cribb comment that the social world is made up of complex and open systems; there are philosophical problems in specifying the relationships between the 'elements' of the social world. However, the majority of good social scientists are aware of these challenges and their methodologies designed to minimise these difficulties. There are also empirical claims and assumptions built into bioethics and it is better to consider the rigour on which they are built and to be critically reflexive about the theoretical frameworks of social description they rest upon, than to leave them unexamined.

Given that bioethics is a 'real world' enterprise we consider it non-controversial that empirical data and research should be[RATHER THAN IS] used as much as possible to enhance argument and to accurately see how problems are constructed. Because the methodologies, and, indeed, the working practices, of social scientists and moral philosophers are very different, (not least because of the difficulties of being prescriptive when one is only used to being descriptive and vice versa), we would urge collaboration between moral philosophers and social scientists when using or undertaking empirical work in bioethics. Our conclusions regarding methodologies in empirical research are documented in our 'Recommendations for the use of empirical data in bioethics' (see section 9). An aspect which is related but which needs to be addressed separately is the issue of the discipline of bioethics itself. What are the goals of the discipline, what is its scope and who should lead the debate?

Bioethics – The Discipline

There is a healthy debate about what bioethics is: what we mean by the term; what it includes or excludes, and the proper methodology for its pursuit. In general, moral philosophers see the methodology of moral philosophy as central and essential to bioethics. Bioethicists from different disciplines do not always share this view. Bennett and Cribb propose two models of bioethics which serve to clarify terminological disputes.

Model 1: Bioethics

The first model of bioethics looks at the philosopher's claim to primacy in bioethics. This is mainly based on the tradition of applied ethics and medical ethics, and, more broadly, moral philosophy which, in order to be objective, must be abstract. Bioethics is therefore seen as a branch of applied ethics – a natural extension of an existing discipline within philosophy. The methodology of applied ethics involves the application of the principles and methods of moral philosophy to practical problems. It is acknowledged and accepted by most moral philosophers that bioethics is a multi-disciplinary mode of enquiry with representatives from medicine, life sciences, religion, economics, psychology, sociology, and others, having an impact and a voice. However, while other disciplines are involved, model 1 bioethics has moral philosophical enquiry as the central method of bioethics. Bennett and Cribb use a quote from Ronald Green to exemplify this approach:

'While ethics and moral philosophy may sometimes represent a relatively small part of the actual work of bioethics, they form in a sense the confluence to which all the larger and smaller tributaries lead, and, more than any other single approach, the methods of ethics and philosophy remain indispensable to this domain of enquiry.'[4]

If bioethics is a branch of applied ethics, which is a branch of moral philosophy, then it is clear why philosophers see their methodology at the centre. This belief is compounded by the difference, and incompatibility, of the disciplines of social science. Philosophy necessarily leaves 'the world as it is' out of its deliberations, when all other disciplines are anchored in practice.

Model 2: Bioethics

On the second model bioethics is truly a multi-disciplinary area of enquiry where sociologists, lawyers, health care professionals, life scientists, philosophers, anthropologists and others use their own methodologies to approach ethics. This model favours collaboration to assess and comment on the social and public policy consequences of bioethical issues.

Problems with Model 1

Bennett and Cribb conclude that debate about the relevance of empirical research to bioethics is partly caused by a confusion between these two models. With model 1, problems in bioethics cannot be solved by appeal to or use of empirical evidence. Moral problems can only be solved by the use of reason and argument and, whilst empirical work may be interesting, it cannot provide moral answers.

For this reason model 1 bioethics is treated sceptically by those who are interested in and believe in the value of, empirical research. Bennett and Cribb comment, for instance, on the Wellcome Trust's perception of model 1 bioethics as 'an abstract exercise carried on over sherry in the tutorial rooms of academic ivory towers'.

There is also scepticism about whether philosophy can resolve moral problems. There is a postmodern view of morality as a personal, subjective and culturally determined phenomenon, and if one is inclined to this view, one person's beliefs are as good as another. If this view is the correct one, it would not make sense to rationally defend moral beliefs, in the same way that it does not make sense to rationally defend matters of taste. But James Rachels make the point that it is the role of reason that distinguishes personal taste from moral judgment:

'In this way moral judgments are different from mere expressions of personal taste – If someone says 'I like coffee' he does not need to have a reason – he is merely making a statement about himself and nothing more. There is no such thing as 'rationally defending' one's like or dislike of coffee, and so there is no arguing about it. So long as he is accurately reporting his tastes, what he says must be true. Moreover, there is no implication that anyone else should feel the same way; if everyone else in the world hates coffee, if doesn't matter. On the other hand, if someone says that something is *morally wrong*, he does need reasons, and if his reasons are sound, other people must acknowledge their force. But if he has no good reasons for what he says, he is just making noise and we need to pay him no attention' [5]

Cultural relativism is, of course, not subscribed to by most moral philosophers who believe that there exist universal truths in ethics or at least that there is an independent standard by which cultural differences can be judged. It is sometimes argued, and Bennett and Cribb do so, that in practice moral philosophers engaged in bioethics are not mere 'sherry sippers'. They *do* still hold that empirical research sheds no light on what is morally right but they also believe their role is not simply to prescribe what is 'right' but to help determine what society's response to moral dilemmas should be. The rightness and wrongness of things deliberated on – through reason and argument – may not necessarily be the only factor that will drive policy.

This is exemplified by the fact that many moral philosophers hold that the human fetus has no moral status or interests, but few of them would suggest that this provides sufficient grounds to use human fetuses as earrings or soup stock, as this would cause distress for no great gain. But of course, this conclusion does not take philosophical expertise to arrive at. Consequentialist thought of this kind almost demands that empirical considerations are incorporated into, and delimit the policy-related musings of moral philosophy. For instance, although reckless transmission of HIV is clearly wrong, criminalising the action may not be the appropriate action if it is shown that legislation is likely to prove counter productive in terms of reducing infection.

It is empirical evidence that tells us that fetus earrings are distressing or that criminalisation of the reckless transmission of HIV is counterproductive. Without it these are just the unsubstantiated intuitions of philosophers. Still, this is not the end of the story –

it is not the case that empirical evidence just indicates what should be done. There are unexamined normative assumptions within the philosophers' deliberations, such as the assumption that in the HIV case, reducing rates of infection is the prime moral good. This aside, Bennett and Cribb note that some may prefer to reformulate these examples to suggest that the evidence is not simply an extra consideration to add to ethical analysis for policy making, but rather that the evidence of distress or counter-productivity *changes* the ethical analysis and the appropriate ethical judgment. Evidence is not an add-on but an intrinsic part of ethical analysis. However this way of putting things makes no difference to the model 1 position. The central point is that the rightness or wrongness of the policy is determined by a process of philosophical reasoning (which can incorporate evidence) and not by the evidence itself. This is not fully persuasive a) because it does not account adequately for the fact that not all philosophers agree; if 'rightness' and 'wrongness' is something determined purely by reason outside the evidence this surely would not be the case (at least not after all had had sufficient time for deliberation) and; b) there are some not so hidden normative assumptions within the philosopher's reasoning (such as the primacy of reducing infection).

The conclusion is that while empirical research may not provide the answers to what is morally right or wrong, it is important when moral philosophers *apply* their analysis of morality to policy-making. If a study existed which showed that criminalisation was a deterrent in seeking treatment to those who suspect they are HIV positive, then this would be a relevant factor for those who hold to model 1 bioethics.

The Wider Value of Social Science

There is a wider value to social science than just selective use of good empirical data or the undertaking of conscientious quantitative and qualitative research, or even than an understanding of the theories and frameworks supporting social science work can provide. The wider value comes from having a social scientific perspective on the world, which is the simple understanding that all human endeavour is value laden and as such, the right questions must be asked by bioethicists.

In order to gain further understanding of what the shortcomings of a moral philosophy-led bioethics might be, the EMPIRE project undertook extensive research into the criticisms directed at bioethics in general and moral philosophical bioethics in particular. It was hoped that analysis of the criticisms, many of which dealt with the lack of use of empirical data, would illuminate the relationship between social science and moral philosophy and the wider goals of bioethics as a discipline.

We found much support for moral philosophers as bioethicists in the general acknowledgement that we need individuals who are trained in the employment of objective and scrupulous reasoning. Moral philosophers fit this criterion; they are regarded as people of good faith, and are already interested in questions of the right and the good and of how we might live well and justly, and hence, they have come to predominate in the field of bioethics. However, whilst moral philosophers may appear to have the requisite skills for tackling these issues, they have also been heavily criticised for what many commentators see as their ahistorical, asocial and acultural approach to bioethical problems.

There are many criticisms leveled at the domination of moral philosophers in bioethics. Most of these criticisms come from social scientists and bioethicists outside the main philosophical tradition. One central criticism is that philosophical bioethicists tend not to question the supremacy of natural science as the only truly valid generator of knowledge. Perhaps one of the most important characteristics of sociology and science and technology studies is their scepticism about the possibility of maintaining the fact/value distinction. As far as sociologists are concerned, the practice of bioscience and technology is necessarily value laden. For example, the bioethicists' acceptance of the authority of medicine is demonstrated in current debates about behavioural genetics. We frequently hear pronouncements from the scientific community that, given enough time and money, genetics will enable us to eradicate a range of behavioural "problems" such as gayness, obesity, violent tendencies, alcoholism, and general anti-social behaviour. The bioethical debate in this instance centres upon the rights and wrongs of using genetic technologies in this way, - namely the pros and cons of manipulating the human gene pool so as to alter behavioural traits, instead of questioning the socially constructed nature of many of these problems in the first place. In other words what is frequently overlooked in these debates is that what society considers "acceptable or problematic behaviour", changes both historically and culturally, and is ultimately defined by both political and economic interests.

The methodology of philosophy also comes in for substantial criticism, particularly directed against bioethics based in the Anglo-American analytic tradition. Firstly, it is claimed that it overlooks the fact that the questions that are so hotly debated in bioethics are not necessarily hypothetical in origin, as in traditional moral philosophy, but arise within the world, and thus, are generated by particular social and cultural factors and influenced by both political and economic interests. Secondly, those philosophers who practice bioethics, in contrast to traditional moral philosophy, are increasingly involved in advising policy makers, and as such, their deliberations and conclusions have a direct impact upon the world.

The abstract nature of philosophy's methodology is the most frequent criticism leveled at bioethics. To achieve the clarity and consistency required of moral philosophy, broadly accepted analytic frameworks and principles such as those of informed consent and respect for autonomy are used within bioethics to assess the acceptability of particular decisions and actions. These, supported by the meta-ethical foundations of consequentialism, deontology and virtue ethics, allow moral philosophers to explore moral dilemmas from a range of perspectives, whilst maintaining objectivity in order that deliberation does not lapse into opinion and preference. However, the flip side of this is that the sparse nature of these analytic frameworks can blind bioethicists to the contextual components that generate moral dilemmas in the first place. For example, whilst a preference for autonomy may be laudable, it ignores the fact that the ideal of "free choice" is socially constructed and situated, and its practice constrained by social, political and economic factors, for example: the values inherent within medicine as a discipline; deference to expert knowledge; class differences; ethnicity; education; economic circumstances; religious factors; work needs and the power and influence of the medical profession.

It is considered that the real value of adopting a more empirical perspective within bioethics lies not in the adoption of empirical methods per se, but in the adoption of a different type of world-view. Moral philosophers as bioethicists need cultivate an interest in, and understanding of, the relationship between structure and agency and make the connection between economic and political conditions and behaviour.

Ultimately, there is a case to be made for philosophers in bioethics to adopt a more sociological perspective. Sociology, from which much of the relevant empirical data derives, could show bioethicists how social structures, historical shapings, cultural settings, power structures and social interaction influence their work. It would allow them to see how bioethics, as a discipline, is constrained by 'disciplinary habits, professional relationships, cultural 'ways of seeing', institutional needs, economic demands, arrangements of power and prestige' and the professionalisation of the discipline within the medical marketplace. A historic reading and teaching of bioethics would also demonstrate how autonomy rather than, for example, justice, came to be the prime moral good. The adoption of a more sociological world-view would help bioethicists to understand why only some issues (generally those involving biotech and pharmaceutical interests) are of interest to bioethics and why clinically-based bioethics is still in the thrall of scientific knowledge as the only valid form of knowledge. Careful reasoning and rigorous analysis require full understanding of the context and circumstances of bioethical debate. The type of expert knowledge we need to be moral experts is truly multi-disciplinary. James Lindemann Nelson sums this up as follows:

'In the end, the contribution of the social sciences to bioethics may be to complete a trend in this interdisciplinary field – that is, to tear down the hold on our imaginations of the notion that the world breaks down along just the lines that intellectual disciplines do'. If bioethics could model a way of understanding and resolving human problems that exhibited both a deep engagement with intellectual disciplines and a freedom from their territorial constraints, that might be among its most profound contribution to its culture'.[6]

Moral philosophers are indeed, correct to say that the world 'as it is' tells us nothing about the world 'as it should be'. However, philosophical bioethicists do not inhabit some hypothetical universe, they exist within the "real" world, they are shaped by, and shape, its institutions. The problems they tackle arise within the world and the decisions they make impact upon it – for bioethics to be effective as a social good, the complexities of this world must not only be acknowledged, but also embraced.

The Problem of Moral Relativism in the Social Sciences

If a rapprochement between bioethics and social sciences is to take place it is, however, also necessary for the social scientists to critically analyse the implicit and explicit moral relativism that characterises many social science disciplines. This is necessary for two reasons: a) whereas an empirical moral relativism seems to be true: different societies have different moral rules, there are insurmountable philosophical problems in maintaining a thoroughgoing normative moral relativism (it would for instance make a mockery of the idea of universal human rights) and ; b) it is constitutive of bioethics as a discipline that it aims at making valid normative claims on a range of levels, not only claims where the validity is limited to one specific society or social grouping. A moral relativist social science cannot therefore be part of the core of bioethics, although it may still contribute to the bioethical enterprise in some ways.

4. The Relevance of Empirical Data and Research for European and Universal Ethical Values and Ethical Theory

It should be uncontroversial that both ethical theory (understood broadly) and empirical research (of some sort) have roles to play in applied ethics and thereby in the derivation and justification of our foundational ethical values. Whatever our view of the exact logical structure of the practical syllogisms or other practical arguments that lead us to ethical conclusions, it is indisputable that these must in most cases contain both ethical and empirical premises, and that these premises must come from somewhere. To reach an ethical conclusion relevant to practical action in the world, we need to know something about ethics and we need to know something about the world.

This is emphasised by even the briefest reading of ethical literature. Such a reading reveals that the literature is full of references to what people want, what they think, and what effects various public policies will have, as well as huge amounts of references to scientific findings and these seem *a priori* all to be empirical matters.

But there are still two areas in which there is considerable theoretical and meta-theoretical disagreement.

The first of these areas of contention concerns exactly what role ethical theory and empirical research should play in ethics. In the first part of this section we will mainly be interested in the relation between the theoretical and the empirical. We will briefly outline four common models for the role of ethical theory in ethics and analyse what role empirical research can have within each of these models. We will especially focus on the degree to which empirical knowledge/beliefs may influence ethical theory and foundational ethical values. We will show that all the models leave a significant space for a two-way interaction between the theoretical and the empirical domain, whether or not that is acknowledged within the model itself. We will further show that there are residual structural problems in getting this interaction to work in the models that acknowledge its existence.

The second contentious issue centres on what kind of empirical knowledge we need. Is targeted research necessary or can we base our conclusions on other kinds of knowledge? This will be the topic of the second main part of the section.

Four Models of Applied Ethics Methodology

Model 1 – The Engineering Model [7]

In the simplest possible model of applied ethics, ethical theory (or some derivative of ethical theory) provides an algorithm that, when applied to a well formed moral problem, can provide the correct moral solution. This may either be in the form of concrete action guidance in the case where the problem concerns a specific, singular action, or in the form of lower level principles covering a class of actions. We can represent this schematically in the following way:

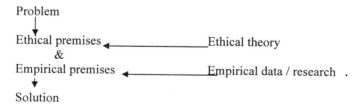

This model can be refined to lead to two variant versions:

Model 1a – The Less Ambitious Engineering Model

This first variant simply acknowledges that not all moral problems may be solved by pure application of theory. There may, for instance, be problems where several moral principles are in conflict and where there is no principled way to resolve such conflict. Simplified variants of the four principles approach exemplify this kind of less ambitious model. [8]

Model 1b – The Regularising Engineering Model

This is perhaps a more truthful version of the engineering model. It acknowledges that it is often necessary to regularise; to re-describe or tame unruly moral problems in order to apply theory 'successfully'. "Irrelevant" features of the problem must be removed or ignored to get to the "core" problem, which has to become sufficiently simple if our ethical theory is to be able to deal with it. This need for regularising and re-description is not specific to the engineering model, but affects most ways of thinking about applied ethics.

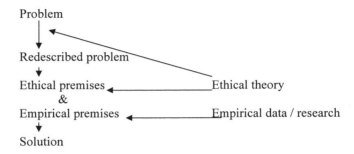

The question of abortion thus becomes reduced to whether or not the foetus has full moral status, or the question of resource allocation in health care to maximising health related quality of life per unit of money spent. In this way we consider the problem under the condition "everything else being equal", but sometimes write our conclusions as if they had been reached "all things considered".

It is worth noting that since most uses of the engineering model are of this last type, it is no mystery that we cannot achieve consensus on solutions, since each ethical theory gives us a different description of the problem!

In the engineering model the relevant empirical premises will differ according to the specific ethical theory that is applied. If the theory is consequentialist then the outcome of each of the possible actions has to be enumerated and weighed before we can decide which action leads to the best consequences, and is therefore to be preferred. Since the consequences and their specific value to people are happenings in the world, the consequentialist analysis will need specific empirical information to get off the ground. As Birnbacher points out, the empirical contribution in deontological reasoning seems to be different and less important [9]:

"It is characteristic of deontological principles to leave much less room for considerations of empirical adequacy and efficiency than consequentialist ones. The reason is that in the process of subsuming individual cases under these principles, deontological principles stand in need of a *semantic* interpretation, whereas consequentialist principles stand in need of both a *semantic* and an *empirical* interpretation. Once the exact meanings of the terms of a principle are fixed, a deontological principle determines unambiguously what is to be done or not to be done in relevant situations. [...] A deontological prohibition to kill another human being simply leaves no room for empirical considerations of prospects and probabilities in the way a consequentialist principle of maximising happiness does. There is no logical gap between ethical principles and concrete rules of action which might have to be filled by empirical considerations."[10]

This analysis is correct as long as we consider the application to a single action of a fully specified deontological system with internal, specified methods for resolving conflict between principles. But if we are not in this situation, empirical considerations may again become important.

Applied ethics does not only reach conclusions and advise on single actions, but also on appropriate regulation or policy in specific areas (e.g. debates about the legalisation and regulation of euthanasia and assisted suicide). In such a situation strategic considerations become important (as Birnbacher points out) and these often require empirical input. If one of the purposes of our regulations is to minimise the number of ethically problematic acts, then it is not sufficient to be able to identify such acts with certainty through our deontological principles. We may also need to know why, how and in what context these acts are performed, and how different regulatory approaches will affect the commission of such acts.

It is also questionable whether an absolute degree of semantic certainty, the fixed meanings of the terms in our ethical principles, can be reached through purely philosophical analysis or argument in any natural language; and if a deontological framework has to use an artificial, meta-language to attain the necessary precision it becomes questionable to what degree conclusions reached by its "application" to real world problems have any validity.

In all the variants of the engineering model ethical theory has supremacy, and there is no direct way in which empirical research can influence ethical theory and foundational ethical values. The communication is purely one way. Ethical theory tells us what empirical premises we need, and it is the role of empirical investigations (scientific and non-scientific) to provide these premises. It is from the explicit or implied perspective of this model that the most simplified critique of empirical research in bioethics often comes. It takes the form of remarks like: "Why should we be interested in what people think? Whether or not X is wrong does not depend on what people think about X!"

If we look more closely at the engineering model it is, nevertheless, often the case that the way the problem is described and re-described involves a large number of (rarely acknowledged) empirical assumptions and that even foundational values in our ethical theories may be considerably influenced by the empirical. As an example, the claim to a very broad right to "reproductive freedom" has become very popular lately, often based on the assertion that reproduction is very important to people. A claim about the importance of a complex social activity like reproduction (we are after all talking about the reproduction of human persons, not of single celled paramecia) cannot be conceptual; it must be at least partly empirical.

Model 2 – Wide Reflective Equilibrium

This model, as developed by Rawls and Daniels, urges us to seek coherence between three different domains 1) moral intuitions (or in Daniels' version considered moral judgements [11]), 2) ethical theory and 3) background beliefs. Elements in each of these can be revised if they conflict with elements in the others in a specific case. There is thus no automatic primacy given to ethical theory.

Here there seems to be a role for empirical research in generating our background beliefs and perhaps also in the process of reaching considered moral judgements and deriving foundational ethical values. The results of such research can impact deeply on ethical theory. The wide reflective equilibrium model thus appears to offer a constructive role for empirical research.

However, this may to some extent be mere appearance. Because ethical theory consists of an already ordered, logically connected (and presumably coherent) set of considerations, any change in one element of the ethical theory domain is likely to require changes in many other elements, and it will thus often be easier to achieve coherence across all the three domains by changing an element in one of the less ordered domains. To use a metaphor from Lakatos' philosophy of science, ethical theory is in reality much more likely to be the "protected core" than any of the other domains, which are likely to become "the protective belt" that deflects all challenges to the core.

Model 3 – Rational Intuitionism

In most writing in bioethics there is no explicit mention of any type of ethical theory or any direct justification of foundational values, and little direct evidence that the writer holds a specific ethical theory [12]. Matti Häyry has described a prominent form of this type of argumentative strategy as rational intuitionism:

"All moral choices and principles should be justified by rational arguments. These arguments, in their turn, must ultimately be based either on the normal usage of language, the cool consideration of rational preferences, or on thought experiments involving hypothetical, and often exceedingly fanciful, examples" [13]

The goal here, as in Model 2, is to achieve global consistency within our complete set of beliefs, moral and non-moral alike. There is, however no specification of domains and all beliefs are, in principle, equal and equally liable to be modified. Again this kind of reasoning should be open to the role of empirical research; it is, for instance, difficult to see how "the cool consideration of rational preferences" can take place in an empirical vacuum.

The introduction of, as Häyry calls them, "exceedingly fanciful" thought experiments points, nevertheless, to a problem in much of the work by "rational intuitionists" that negates the seeming openness to empirical knowledge. Many of these thought experiments involve counterfactuals, and some of these counterfactuals involve situations that are either currently scientifically impossible, or they ask us to imagine societal configurations that are far from our current ones and may never have been instantiated anywhere. It is often assumed that social beliefs about; attitudes towards, or actions with regard to, specific phenomena are 1) easily malleable, and 2) will actually change in the desired direction. Thought experiments have an important role in ethical reasoning, but if used too often and too liberally they may entice us to forget that they are only thought experiments, and that there is a current society, with a specific set of structures in which our moral analysis has to work. It may be that we can get anywhere from where we are now (although that is in itself questionable), but there is no reason to believe that our most immediate social successors will not share a great deal with us. The use of thought experiments may lead us to systematically downplay the role of empirical reality in favour of a preferred, imagined, reality.

Model 4 – Local Coherentism

The model we have just discussed sought global coherence in the relevant set of moral judgements and beliefs, but it is possible to be less ambitious and just seek local coherence within some specific area of ethical interest, e.g. reproductive ethics. This kind of local coherentism is also quite common in articles about bioethical issues. The basic argumentative structure is often some kind of parity of reasoning argument where it is shown that if we argue in a specific way in one area, say about abortion, we thereby commit ourselves to certain conclusions in other areas, say, about embryo experimentation.

Alternatively, an attempt is made to find one principle that can explain our moral judgements in a certain area, for instance, personhood approaches to the ethics of killing. Local coherentism is problematic because it 1) allows the writer to hide unpleasant consequences of his or her position as long as they occur outside the area discussed, and 2) assumes that local coherence is always a positive achievement, even though it may sometimes be achieved at the expense of global coherence.

Evidence-based Ethics

Even if it is indisputable that empirical premises are indispensable features of many ethical arguments, there can still be disagreement about what kind of empirical knowledge is needed. Is personal experience or anecdote a sufficient basis or is something different (stronger?) needed?

In many other areas of practically important decision-making we are moving towards a standard of evidence-based decision-making, which, in its simplest form, is the idea that important decisions affecting other people should only be taken on the basis of the best available evidence, and that good quality evidence should be developed in those areas where there is none at present. The most prominent example is evidence-based medicine [14], but similar approaches can be seen in other fields like education or environmental interventions. In still other areas the recognition of a need for a solid evidence base are so old that we may forget about it, but think for a moment about what attitudes we would have if somebody advocated engineering bridges based on pure theory.

It is very difficult to mount an argument that ethical decision-making should be exempt from being evidence-based. After all, we often claim that our ethical decisions are "all things considered" decisions. That seems to require that what is known about relevant facts have been taken into account, and that we acknowledge that our conclusions may be weak if some of the "facts" are disputable.

For instance, if we analyse whether patients should be given full information about treatment options and their effects and side-effects, what happens when patients are given full information and what happens when they are not, must be highly relevant. Just theorising about the amount of distress caused by full disclosure, or recounting a few choice stories seems woefully inadequate when we can study what actually happens.

But could we not claim that no specific empirical research is needed, that we can rely on "what is commonly known" about the empirical premises in our arguments? Firstly, it may of course be the case that nothing is commonly known, because these particular empirical questions only become interesting in the context of our ethical argument. Secondly, the authority, or epistemic warrant of what is commonly known is often questionable. Most bioethicists would recoil in horror if it was suggested that ethical premises in arguments should be based on "what is commonly known about morality". Why should we react differently concerning other items of common knowledge?

Another possible claim specifically against the use of facts generated by research in the social world is the claim that data generated by social research are not pure, they carry a theoretical baggage and are therefore in themselves already normative. This may well be true, we would not necessarily expect a Marxist and a Libertarian sociologist studying power relations in the operating theatre to reach the same conclusions. But this is no more true of research-based knowledge about society than it is about any other kind of knowledge about society. Our personal beliefs about our society are heavily influenced by folk-theory and by watered down versions of currently popular social theorising and this influence is often implicit and opaque to the holders of these beliefs. Social science generated knowledge is often preferable because there is an explicit acknowledgement of the theoretical background.

Finally, the philosophical user of empirical data might raise the following complaint: "You have completely misunderstood me. When I talk about what people think or desire I am not drawing on any kind of empirical psychology or sociology, I am drawing on philosophical psychology, which is something completely different. The "philosophical psychologist" does not need any empirical data to work on. It is all conceptual."

It is undoubtedly true that empirical psychology and philosophical psychology are two very different enterprises, but it's also undoubtedly false that philosophical psychology can expect to gain any deeper positive understanding of the psychology of human beings without some empirical input. A Rylean de-construction of the ghost in the machine is good as far as it goes, but it does not, for instance give us any positive insights into the phenomenon of human consciousness [15], and philosophical analyses of rationality do not exhaust the space of reasonable decision-making.

Conclusion

In this section we have argued for the following conclusions:

1. That application of ethical theory and the derivation of foundational ethical values necessarily involve empirical input;

2. That the empirical input does not only describe "facts" about the world, but that it also has a function in our initial regularising of ethical problems and in the specification of vague terms in our ethical theories;

3. That there may be an inherent tendency to resist changes to ethical theory even in approaches that should be more open to a role for the empirical;

4. That the strength of the epistemic warrant for our empirical claims affects the strength of our final ethical conclusions. We should therefore seek the best possible evidence.

5. The Use of Empirical Data in Ethical Argument and Regulation

There are many examples of empirical data which have had an impact on ethical thinking. Some, such as Stanley Milgram's experiments on obedience are notorious for their conclusions about human nature, some controversial for their examination of gender difference such as in Carol Gilligan's 'Ethic of Care'. Others have a direct impact on policy, for example, by indicating how necessary bioethics is in regard to the humane and equal treatment of citizens regardless of ethnicity, the Tuskegee Syphilis study in the U.S being one of the more notorious examples of non-therapeutic experimentation on human subjects. In his work 'The Gift Relationship', which combined an empirical study and advocacy of altruism, Richard Titmuss helped save the U.K. health service from privatisation. We analysed these important examples of empirical work to determine where the importance lay in terms of impact on bioethical thinking. This section provides a summary of the cases and their most important contributions.

Carol Gilligan's 'Ethic of Care'

Carol Gilligan published 'In a Different Voice' in 1982. Her empirical claim, based on qualitative research on a small population of America women, was that men and women have fundamentally different approaches to morality. The male approach to morality is that individuals have certain basic rights, and that you have to respect the rights of others, so morality imposes restrictions on what you can do. The female approach to morality is that people have responsibilities towards others, so morality is an imperative to care for others. In other words, where men focused on justice, women thought about ethics in terms of care, responsibility and interdependence in relationships. Gilligan's research was a response to the work of Lawrence Kohlberg, whose study showed that girls rank lower in their moral development than boys.

The study has been criticised for the smallness of the study group (144 participants), despite the qualitative work being fairly intensive. Gilligan emphasised the fact that her interviewees were from different social and cultural backgrounds but they were still all products of American society and its ideas and values [16].There is the additional problem of whether it is possible to deduce anything reliable about people's moral beliefs from what they *say* they think. Furthermore, rival theories exist which explain the difference between men and women through the possibility that powerless groups often learn empathy because they need the protection of others, rather than through some inherent difference. Gilligan's work can, of course, be criticised for its methodological shortcomings (a condition of all empirical work), but its use to bioethics must be partly in the controversy it generated, the different ways of thinking it encouraged and its, perhaps unintended, effect of showing aspects of morality as politically and culturally situated.

The Milgram Experiments

Stanley Milgram's studies in the early sixties were intended to examine the conflict between obedience to authority and personal conscience. In the research, 'teachers', who were the subjects of the experiment, were asked to administer a series of electric shocks, of increasing intensity, on another person. The rationale given was that the experiment was to analyse the effect of 'punishment' for incorrect responses on learning behaviour. Sixty percent of the 'teachers' obeyed orders to punish the learner up to the 450 volt maximum. No subject stopped before 300 volts. When the experimenter was questioned by concerned teachers they were told the experimenter assumed 'full responsibility' and this response was generally accepted.

The relevance of this research to applied ethics is as an exemplar in discussions about research ethics, both with regard to the ethics of deception, and with regard to a pure consequentialist justification of research which is harmful to subjects but produces knowledge of great value. The research is so provocative and the findings so interesting that it is difficult to find better exemplars around which to centre research ethics discussions. The more interesting question is, however, what role Milgram's findings will have in future ethical or regulatory arguments. The most obvious role is in arguments about the design of hierarchical systems and the regulation of action within such systems. If we are interested in preventing ethically problematic actions, then we should pay attention to Milgram's findings. The obedience experiments shows that the social context in which a given form of authority is used may be much more important in determining the actions of subordinates than the individual subordinate's personality or ethical considerations. It may often be more useful and efficient to restructure institutions than to focus on individuals and individual responsibility. The experimental obedience literature contains a wealth of information about the contextual factors that influence obedience, and this information is clearly relevant in designing 'ethical interventions' of various kinds.

Another use of Milgram's findings is in retrospective analysis of ethical failure. There is often a tendency when analysing situations where ethical failure has occurred to focus almost exclusively on the responsibility of the individuals involved. But this focus will often be misdirected and will exemplify the attribution error that leads people to grossly underestimate the level of obedience displayed in the Milgram experiments and to overestimate the differences in personality between obedient and defiant subjects. It may well be that our analysis shows that a given actor ought to have acted in a different way than he or she did, but before we apportion blame for this ethical failure we should take time to consider whether the average person in possession of all the ethically relevant facts would have acted differently in the same situation.

A third possible use is in discussions about the kind of moral personality we ought to cultivate in ourselves and others. The obedience experiments seem to indicate that under certain fairly common circumstances it is not enough to have the cognitive knowledge that an act is wrong, and to have the requisite emotional responses in order to act rightly (many of Milgram's subjects presumably both knew and felt that they were doing something wrong). Something more is necessary and this something may be totally unrelated to those character traits that we have traditionally called virtues. There are research findings indicating that for instance, high social intelligence (the ability to understand the thought, feelings and intentions of other people), significantly promotes defiant behaviour in Milgram type experiments.

The Tuskegee Syphilis Study

From 1932 – 1972 the US Public Health Service conducted an experiment involving the observation of the course of untreated (and under-treated) late syphilis in 400 African American share croppers in Alabama. The intention was to compare the prognosis of the untreated syphilis participants with that of another, non-syphilitic population. At the start of the study the best available treatment for syphilis was 'salvarsan'. However, it was noted that, particularly in relation to late syphilis, the symptoms of the disease seemed less severe that the side effects of the treatment. The rationale behind the Tuskegee study was an extremely sensible one. The population in question had an extremely high rate of syphilis. They were underprivileged and had therefore received little or no treatment which was important if the study was to reveal scientifically useful data.

Tuskegee is cited a prime example of unethical research involving deliberate non-treatment, under-treatment, deception and exploitation. The New York Times reported it as the longest non-therapeutic experiment on human beings in medical history. The study has been labeled racist and an example of genocide. Yet despite its many ethical problems, the example of Tuskegee is sometimes unfairly and potentially destructively employed. For example, a distinction must be drawn between the beginning of the project, when guidelines were not in place to prevent practices now considered unethical, and the continuation of the project after the Nuremburg Declaration and the Declaration of Helsinki. Implications for research ethics are grouped under three ethical criteria used by research ethics committees in the UK today, namely scientific validity, participant welfare and participant dignity.

For its time, the observational method would have been considered perfectly adequate. There was no clearly defined protocol at the start of the study which, again, was largely a product of the times. The results gained were useful and the study is still quoted. Criticism is largely directed at the lack of consent. Information was sparse and deception was used to persuade participants to undergo diagnostic spinal taps. Though we might accept lower standards in 1930, the US Public Health Service cannot reasonably be excused after the Declarations of Nuremburg and Helsinki.

The legacy of Tuskegee is that it has contributed to the formulation of research ethics principles including stringent patient (informed and voluntary) consent guidelines; a right to the best proven treatment and; an expanded role for nurses in obtaining informed consent (refusal to rely on physician's orders or to disregard ethical standards). The exposure of the scandal and subsequent public condemnation led to the creation of the National Commission for the Protection of Human Subjects of Biomedical and Behavioural Research and the National Research Act (requiring the establishment of Institutional Review Boards) in 1974. This ensured that research involving 'vulnerable participants' would be given extensive review prior to being approved. Finally, a Presidential apology in 1997, led to planning and funding of the Tuskegee University National Center for Bioethics in Research and Health Care, which was established in 1999 and is the only ethics centre funded by the federal government. This Centre is dedicated to community outreach and research to look into and redress medical mistreatment of African Americans.

Richard Titmuss' 'The Gift Relationship'

'The Gift Relationship' was published in 1970. It compared the system of voluntary blood donation in the UK with the US which had a joint system of donation and sales. Titmuss concluded that the voluntary system was better across the board. The quality was superior and it was more efficient, leading to a safer system all round. Its impact reached the US where the government, after consultation with Titmuss, attempted to stimulate voluntary donation and mandated the labeling of blood from paid donors. Although it was an empirical work, Titmuss's most important conclusions were about the importance of altruism. He argued that privatised and market driven healthcare is in tension with the moral importance and value of altruism within society. When blood becomes a commodity its quality is corrupted (blood in the US was four times more likely to infect recipients with hepatitis than blood in the UK).

The 3810 donors in Titmuss's survey used a "moral vocabulary" to explain their reasons for giving. This vocabulary appealed to notions of the "universal stranger" as someone who might be in need and deserved to be helped. People gave blood out of a feeling of general social obligation. They did not give blood only because they or their families might need it. The empirical work, completed in the late sixties, can be criticised for its methodological failures as research methods have come a long way since then. It was also argued that the flaw in the U.S. system was a product of the wrong system of liability, but that the purchase and sale of blood was still an acceptable method of procurement.

The Legacy of Empirical Work

All the above examples had a far-reaching impact on bioethics and moral discussion in general. All of these studies are well known outside the field of bioethics and have impacted the policy process in some way. All of them are methodologically flawed. Their methodological shortcomings do not detract from the educational value they hold or the controversy they have generated. Work on explaining gender and race differences, on trying to articulate our obligations to others and in trying to understand the limits and constraint on our morality have nourished the conceptual side of bioethics.

While generating debate and controversy is valuable to the field, what was very interesting about our analyses of these works was that in each case the complexity and detail of the research was subsumed under the main thesis of the research. When they are used to support bioethical debate they have effectively been reduced to what they most symbolise. Titmuss is a byword for the value of altruism; Gilligan represents gender difference in approaches to morality; and this simplified presentation means the nuances of the research and their empirical value are lost to a single message.

This is difficult to counter and stems from the tendency for debates to polarize. Instead of trying to achieve something like wide reflective equilibrium in ethical debate, different 'sides' use the single main message from previous empirical work to bolster their positions. For example, Titmuss is regularly used in debates about organ selling when it is arguable whether his work has any relevance at all to the debate.

Despite the inherent problems in empirical work from flawed methodology to reductionism, it makes an invaluable and absolutely necessary contribution to the bioethical debate. Even strict moral philosophers must take empirical claims seriously rather than be accused of being 'content to invent their psychology or anthropology from scratch' [17]. A policy of being open about the flaws in methodology and a willingness to analyse the aspects of empirical work that are broader but more nuanced than the main thesis can only be advantageous to the bioethics enterprise.

6. Methodologies

There are several problems which surround the issue of methodology in the use of empirical data and research in bioethics. Firstly, there is the problem of establishing what empirical information is relevant and also whether the methodology used in generating the information was appropriate (was the information taken from a relevant study population, for instance). The use of empirical data and research in bioethics is further complicated by the fact that philosophical bioethicists generally know very little about research methodology and those who are accomplished in research methodology- social scientists- may know very little about ethics and will generally be unaccustomed to making or analysing normative statements. It is not feasible to expect either the philosopher or the social scientist to learn a further discipline and the best way forward ideally would be collaborative work between social scientists and philosophical bioethicists. There are certain rules of thumb however which can be used as guidance for bioethicists when using empirical data or engaging in empirical research, and these will be discussed in this section. The main conclusions of this chapter comprise the 'Recommendations for the use of empirical data in bioethics'.

Findings from social science and psychological research are often used as premises in ethical arguments and in arguments concerning regulation of medicine and biotechnology. It is important that such findings are used in a responsible and transparent manner. It is also important that empirical research initiated in bioethics, or with the aim of influencing policy-making in the bioethical field is of a high standard, and well thought out before it is initiated.

If empirical data are used as more than illustration or anecdote in ethical arguments it is important to assess their validity and generalisability, and to perform a comprehensive search of the literature to discover to what extent there is intellectual disagreement both with regard to the actual findings, and with regard to the underlying analytical and theoretical frameworks. There may be a temptation to just utilise empirical data which accords with the understanding of the problem held by the researcher. This should obviously be resisted.

As a general guide to embarking on empirical research it is most important that the researcher should be familiar with the research which has previously been done in the relevant area in order that they do not duplicate work. It is also important to take into account the 'grey' literature which would include any reports, consultation documents, Ph.D. theses and government white/green papers.

Social scientists have long understood that the research process is not value free. Even in science, widespread intellectual disagreement indicates that values inherent in the research process have a bearing on the results. Given that there are always external issues involved in the production of research, the following questions should be asked in order that the research can be assessed appropriately. The user of the research should know why the research was commissioned and who funded it. This gives an insight into what results were hoped for or expected. It should be known who conducted the research and which methods were utilised, it may then be possible to determine whether the methodology was appropriate. For example, a piece of empirical research on a relatively small population may not accurately reflect the state of affairs nationally. The users of research should make themselves aware of the general limitations of the methodology used and not just aspects of it such as study size. A good indicator of quality is discussion in the publication itself of how the study results were interpreted and used. Interpretation may be a large part of a study, particularly a qualitative research study. Finally, in the development of the study there may be underlying pre-suppositions such as information about a genetic condition generally being something that is desirable. When undertaking empirical research or using another's research, care should be taken that assumptions are clearly identified and not implicitly built into the research design.

Because philosophical bioethics generally employ abstract methodology, they have historically failed to realise or give due weight to the many factors which have an influence on moral dilemmas. These can include; institutional arrangements, cultural tendencies, economic constraints, religious beliefs, and historic and political factors. For example, it may be that using philosophical analysis alone would allow one to come to an understanding that sex selection is appropriate but in countries where women are discriminated against in comparison to men, the availability of sex selection may cause more problems than it solves. Also, the moral compass of the researcher will not necessarily accord with others as, in general, moral concepts and norms derive their meaning and their force for people from the social and cultural surroundings in which they are embedded. An example of this would be gang culture or organised crime culture where there are rigid and clearly defined moral codes which must be followed for members to remain included. A willingness to understand moral concepts as situated rather than universal will help in understanding types of qualitative empirical research such as ethnography in particular.

Not only should one use the appropriate principles but also bring to the problem an understanding of the constraints of a population and the relevant unequal power structures involved. This means that considerations can be weighed appropriately. Also, bioethicists need to understand to what degree their values are shared by others, and to what degree their values can be argued to be the only acceptable values. For example, what some philosophical bioethicists assume to be a universally desirable condition – such as autonomy, may not be what actors in a health care situation actually prefer, some patients may simply prefer good clinical guidance.

Empirical research does not stand on its own. The researcher should try to understand the relevant theoretical background of particular methodologies. For example, the rationale behind why ethnography may be the best type of methodology for a particular study. It is problematic to simply rely on empirical data without this understanding. Just as it is the responsibility of the bioethicist to consider the particular strengths and weaknesses involved in the methodology of the empirical work being used, so he should be able to understand it in the context of its theoretical underpinning and, again, to appreciate how this may affect conclusions and prescriptions. This will seem like an enormous task to those inexperienced in social science methods but good empirical studies will often clearly outline the study's limitations. If a study fails to the do this it may give rise to concern. It often strengthens validity to take account of the findings of different types of research around the same issue (for example, questionnaires, interviews, observation, experimental studies, etc). Also it is very helpful to consider what different disciplines, such as law, psychology, sociology have had to say about the particular phenomena being considered.

An obvious point but one which is not always adhered to is that research used or cited must accurately reflect the current situation and care should be taken that any empirical research is not 'out of date'. For example, to build present arguments on what doctors thought about abortion 20 years ago would be misleading. Bioethicists cannot just say 'research has shown that...' but need to state the timing, context and size of the study.

For bioethicists doing their own empirical research, we stress that it is dangerous to just 'dip' into another discipline - far more work is required. To do empirical research, the bioethicists must be knowledgeable about different methods in order to determine the most appropriate one for the purpose and also to understand the limitations of the method used in order to consider this in the conclusions reached. Another common problem is the potential to assume that words have the same meaning for patients or populations as for philosophers. For example, a concept such as autonomy is not always understood the same way. Philosophers tend to speak of autonomy as though it was a quality, or characteristic or 'thing' possessed by all competent people when in fact, as we mentioned earlier, the autonomy of anyone at any given time is subject to multiple constraints. This is particularly true for those who are suffering a health care problem or related issue.

Most of these points are directed at philosophical bioethicists. Similar qualifiers need to be made to social scientists engaging in bioethics. They should possess at least a working knowledge of commonly used ethical principles and frameworks, the strengths and weakness of these frameworks, their philosophical justification, as well as an understanding of the goals of bioethics. Furthermore, they need to know what ethical questions to ask in order to make their own research relevant. Finally, in all areas of research, science and social science, there is often a preferred outcome. No one engaging in research, for whatever reason, should suppress data contrary to their own findings or preferred outcome.

Special Issues When Contemplating Regulation

The introduction or change of regulation in an area of biotechnology or medicine often influences peoples' lives in profound ways. It may allow or deny people access to specific treatments, stimulate or hinder certain kinds of research, or allow or prohibit the sale of goods or services. It is therefore particularly important that arguments used to justify and shape regulation are valid, and that their premises, including empirical premises, are true. In the context of regulation it is therefore reasonable to expect a higher standard in the use of research findings and initiation of research projects.

Regulation and control need not necessarily be prescriptive and restrictive. It can be directive and instructive. Modes of regulation and control may include legislation, regulation, administrative directives and guidelines for ethical committees and institutional review boards.

When undertaking research and using empirical data to support regulation, the research must be of a high methodological standard. Furthermore, before research is commissioned it is important to know exactly how the results are intended to be used. If the public is formally consulted it is, for instance, important to know in advance to what extent public opinion will be taken into account. On the issue of public opinion, it is important not to conflate "public opinion" as gathered in polls or assumptions about public opinion, with the considered ethical views of the public which would require in depth consultation.

7. Relevant Evidence for Policy Making

In order to investigate what kinds of empirical research are most relevant to decisions about regulation or control of new technologies or healthcare interventions, we performed a survey of the national ethics committees (or similar bodies) in the EEA countries and five newly associated states. The survey focused on the use that the committees made of empirical research in their deliberations and policy advice.

Of the 20 questionnaires sent out, 15 were returned (75%). The responding committees were contacted by phone or e-mail if any of their responses needed clarification.

Initiation of Empirical Research

The first section of the questionnaire focused on whether the committee in question had commissioned any primary empirical research or any reviews of empirical research during the last five years.

Five of the 15 committees had commissioned new empirical research to fulfill their own perceived needs. Examples of projects were:

a) A high circulation questionnaire surveying participants in clinical drug trials about informed consent procedure and their understanding of it;
b) Focus-group based qualitative sociological research about patients' rights;
c) Research about Human Assisted Reproduction techniques, overseen by a sociologist on the Committee;
d) A questionnaire regarding clinical ethics Committees and hospital decision-making;
e) A study of women's attitudes towards routine pap smears for cervical cancer, carried out by students within the Department of Sociology at the National University;
f) A study of doctors' attitudes to prioritisation within the health care system, conducted by a philosophy student.

Five of the 15 committees had commissioned reviews of empirical research in a specific field, but there was not a complete overlap with the five committees that had commissioned primary research. Two committees had commissioned only reviews, and two had commissioned only primary research. Examples of reviews were:

a) An analysis of scientific papers in National Biomedical journals to establish whether they contain any reference to research ethics;
b) A literature study of the psychological impact of prenatal risk-assessment screening for Downs Syndrome and Neural Tube defects;
c) A literature survey relating to medical decision-making,

d) Literature reviews for Working Parties on such topics as empirical studies of attitudes towards genetic enhancement and prenatal selection for non-disease traits.

Other Sources of Empirical Research

The second section of the questionnaire focused on other sources of empirical research findings.
The committee was asked to indicate on a three point scale ('Frequently", 'Occasionally', 'Never') whether it used or sought empirical information from one of seven sources:

a) Research conducted by members of the organisation (but not funded by your organisation);
b) Research reported in national journals;
c) Research reported in international journals;
d) Reports from the European Commission or other EU institutions;
e) Special reports by other national commissions/ advisory bodies;
f) Special reports by other international commissions/advisory bodies,
g) Other

The results are shown graphically below:

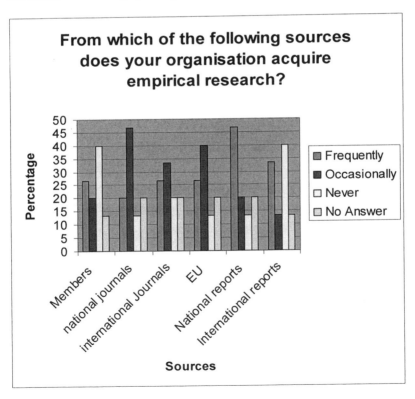

The responding committees were asked to give examples within each category and we have reproduced some of them below:

a) Research conducted by members of the organisation:
-Studies in legal affairs;
-Evaluation of the Population Screening Act;
-Genetics; Medical Law; Medical Ethics;
-Teenage suicide as a medical problem,
-Questionnaire on Euthanasia conducted by the Croatian Medical Chamber 2001

b) Research reported in National Journals:
-All National Scientific Journals;
-Journal of the Swedish Medical Association;
-Daily Press cuttings; 30 National Journals;
-Journal of the Danish Medical Association;
-Special editions on Bioethics in national newspapers;
-Major National Newspapers,
-Journal of the Croatian Medical Association

c) Research reported in International Journals:
-Human Reproduction; Fertility and Sterility;
-All relevant international scientific journals;
-Science; Journal of Medical Ethics; Hastings Report;
-Nature; Science; Bioethics; Hastings Report;
-Bioethics;
-International Academic Reviews;
-CQ of Healthcare Ethics; HEC's Forum,
-Science and Engineering Ethics

d) Special reports from other National Commissions or advisory bodies:
-Nuffield Council; Health General Council UK;
-Royal Netherlands Academy of Science; Royal Netherlands Association of Medicine; Rathenau Institute; Council for Health and Social Services;
-Swedish National Board of Health and Welfare; Swedish Research Council; Governmental and Parliamentary Committees;
-Local Ethics Committee reports;
-Health Ministry Reports;
-German Reference Centre for Ethics in Life Sciences,
-French and Slovenian Commissions

e) Special Reports by other International Commissions/Advisory Bodies
-professional organizations;
-Wellcome Trust; Human Genetics Commission; Consumers Association; Individual Researchers findings

Public Consultation

The final section of the questionnaire focused on public consultation activities and the results are given in tabular from below:

Does your committee	Frequently	Occasionally	Not yet, but plans to do so	Never	No answer
1. hold public meetings to get public input?	1	7	3	3	1
2. commission opinion polls?	0	0	3	10	2
3. hold citizen's juries?	0	0	3	10	2
4. hold consensus conferences?	0	2	3	9	1
5. circulate draft reports for consultation?	2	4	1	7	1

Conclusion

Our survey shows that empirical research plays a role in the work of national ethics committees, but that this role varies significantly from committee to committee. Some committees initiate their own research, whereas most rely on systematic or non-systematic use of already existing research. The sources of already existing research findings also vary considerably between committees.

References

1] See the papers by Holm and Harris in this volume.
2] Bennett, R and Cribb A, (2003), 'The Relevance of Empirical Research to Bioethics' in Häyry, Matti Takala, Tuija and Herissone-Kelly, Peter, (eds), *Scratching the Surface of Bioethics* Amsterdam and New York: Rodopi
3] Richards, Janet Radcliffe *et al* (The International Forum for Transplant Ethics) (1998), 'The Case for Allowing Kidney Sales' 342 *The Lancet* 45
4] Green, Ronald M, (1990) 'Methods in Bioethics', *The Journal of Medicine and Philosophy*, Vol. 15 No. 2, April 1990, p182
5] Rachels, James, (1993),*The Elements of Moral Philosophy* [2nd Edition] New York: McGraw Hill, pp 10-11
6] Nelson, James Lindemann, (2000),'Moral Teachings from Unexpected Quarters' *Hastings Centre Report* Jan-Feb, p 15
7] First named by Caplan in Caplan, AL (1987), 'Can Applied Ethics Be Effective in Health Care and Should It Strive to Be?, in Ackerman, TF, Graber GC, Reynolds CH, Thomasma DC (eds.) *Clinical Medical Ethics – Exploration and Assessment*, Lanham, MD: University Press of America, pp. 131-143
8] Holm, S, (1995), 'Not Just Autonomy - The Principles of American Biomedical Ethics', *Journal of Medical Ethics* 21: 332-38; Holm, S, (1999), 'Principles of Health Care Ethics: Solution or Problem?' in Launis, V, Pietarinen, J, Räikkä, J (eds.), *Genes and Morality – New Essays,* (Value Inquiry Book Series vol. 83) Amsterdam, Rodopi, pp. 51-62
9] A similar point has been made previously by Brody BA (1993), 'Assessing Empirical Research in Bioethics, *Theoretical Medicine* 14 pp. 211-219

[10] Birnbacher, D, (1999) 'Ethics and Social Science: Which Kind of Co-operation?' *Ethical Theory and Moral Practice* 2 pp. 319-336
[11] Daniels, N, (1996), *Justice and Justification: Reflective Equilibrium in Theory and Practice*, New York: Cambridge University Press
[12] The fact that there is no explicit use of ethical theory to be found does not mean that the writer does not have an ethical theory, or has not used an ethical theory to reach the conclusions that are argued for, it only means that the writer thinks a theory-free style is the most effective rhetorical tool.
[13] Häyry, M (2001), *Playing God - Essays on bioethics*, Helsinki: Helsinki University Press
[14] Evidence based medicine often views a rather narrow subset of evidence, i.e. evidence for effectiveness as if that is the only valuable evidence. This is a mistake, but not one that is a necessary component of an insistence on good evidence for decision-making.
[15] A point realised in the new fields of Consciousness Studies and Neurophilosophy
[16] see Gylling's piece in this volume.
[17] Darwall, S, Gibbard, A and Railton, P, (eds) (1997) *Moral Discourse and Practice: Some Philosophical Approaches*. Cited in Doris, J and Stich, SP, 'As a Matter of Fact: Empirical Perspectives on Ethics' to appear in Jackson, F and Smith, M (eds.) *The Oxford Handbook of Contemporary Analytic Philosophy*, UK, Oxford University Press, forthcoming
[18] Empirical Methods in Bioethics – EMPIRE project, funded by the Quality of Life Research Programme of the European Commission (Directorate General Research), Contract Number QLG6-CT-1999-00517

Author Index

Albretsen, Jørgen	119
Dawson, Angus	41
Edgar, Andrew	8
Gammelgaard, Anne	99
Goodacre, John	109
Gylling, Heta Aleksandra	61
Hallowell, Nina	28
Harris, John	18
Holm, Søren	3,119,131
Irving, Louise	28,131
Levitt, Mairi	109
Norup, Michael	76
Øhrstrøm, Peter	119
Solbakk, Jan Helge	53
Takala, Tuija	69
Van Delden, Johannes	89
Van der Scheer, Lieke	89
Van Thiel, Ghislaine	89
Weiner, Kate	109
Widdershoven, Guy	89